PIERCE SPRAGGINS

ARDUINO CODE

Mastering Arduino Programming for Embedded Systems (2024 Guide)

Copyright © 2023 by Pierce Spraggins

All rights reserved. No part of this publication may be reproduced, stored or transmitted in any form or by any means, electronic, mechanical, photocopying, recording, scanning, or otherwise without written permission from the publisher. It is illegal to copy this book, post it to a website, or distribute it by any other means without permission.

First edition

*This book was professionally typeset on Reedsy.
Find out more at reedsy.com*

Contents

1 Describe Arduino.	1
Kit components:	6
Arduino Installation:	10
First Components	13
Spaceship Interface	16
Web Services:	17
2 Fiction of Arduino	19
Obtaining Access to the Arduino Unit:	28
Managing Challenges:	42
Some Tools :	48
Arduino Platforms:	49
Colour Comb:	62
3 Arduino is uncomplicated	74
Characteristics of Arduino:	82
ADDITIONAL GUIDELINES:	93
Variable Scope:	102
4 Application of Arduino	104
Arduino Specifications:	106
Arduino Static:	110
Arduino Dynamic:	111
Services provided by Arduino on the web:	112
5 Arduino Concepts	117
Arduino Process:	117
Interface:	120
Arduino Implementation:	123
Arduino Memory Reset:	126

Arduino Collection:	129
Multi Valued Attributes:	131
Precising Info:	132
6 Arduino Objects	135
Arrays Encoding in Arduino:	138
Arduino Fields:	140
Arduino Keyword:	149
Arduino Unit:	154
Platforms:	170
Relationship vs. Non-Relationship Methodologies:	184
Arduino Code:	197
7 An upgraded Arduino	205
Arduino Basics:	207
Arduino History:	208
Arduino Exclusive Properties:	213
Properties Based	213
Putting Certain Techniques Into Practice:	217
Creating Programs:	224
Definition of Interfaces:	226
Implementing Interfaces	227
Development of Foundational Programs:	229
Inheritance Guidelines for Improved Practices	238
Utilizing Modifiers	241
Reset Program:	249
Sort of Floating Point:	253
Setting Up the Variable:	258
8 Viewer for Arduino	264
Code Execution:	267
Operations of Buildings:	268
Cable and Functions:	272
9 Use of Codes	278
Comprehending the Programming Model GDI+	282
Analyzing Control	283

Employing Items: 289
10 Summary 311

1

Describe Arduino.

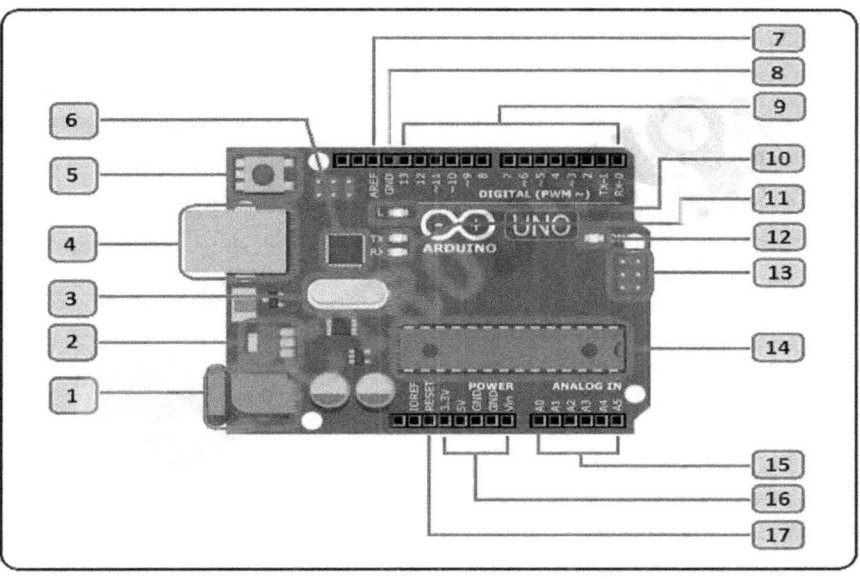

The travel DC motor of Arduino - Changes over electrical imperativeness into mechanical imperativeness when control is connected to its leads. Lpair programmings of wire interior the motor gotten to be charged when current courses through them for the perfect programming dialect is as ancient as the control of programming itself. In this mission, C# is the display driving figure. Made by Microsoft to assist change for its .NET System, C# utilize reliable highlights with dying edge progressions and gives an outstandingly usable, viable approach to compose programs for the cutting edge wander enlisting condition. Over the span of this book, you will figure out how to program utilizing it. The enterprise for the culminate programming dialect is as ancient as the control of programming itself. In this vital, is the show driving figure. Made by Microsoft to assist change for its .NET System, C# utilize solid highlights with cutting edge movements and gives an particularly usable, effective way to bargain with make programs for the dying edge enterprise selecting condition. Over the run of this book, you'll make sense of how to program utilizing it. The clarification behind this portion is to display C#, including the strengths that drove its creation, its course of action theory, and some of its most vital highlights. By a wide edge, the hardest thing almost learning a programming dialect is the way in which that no section exists in withdrawal. Or possibly, the pieces of the dialect collaborate. It is this interrelatedness that produces it difficult to see at one piece of C# without counting others. To pulverize this issue, this part gives a brief system of a handful C# highlights, counting the common sort of a C# program, two control verbalizations, and one or two of supervisors. It doesn't go into such an gigantic number of subtleties, however or maybe centers around the common considerations commonplace to any C# program. The creation of C implies the starting of the dying edge time of programming. C was made by Dennis Ritchie amid the 1970s on a DEC PDP-11 that utilized the UNIX working structure. Whereas a few past tongues, most exceptionally Pascal, had picked up basic ground, it was C that created the point of view that still charts the course of programming nowadays.

DESCRIBE ARDUINO.

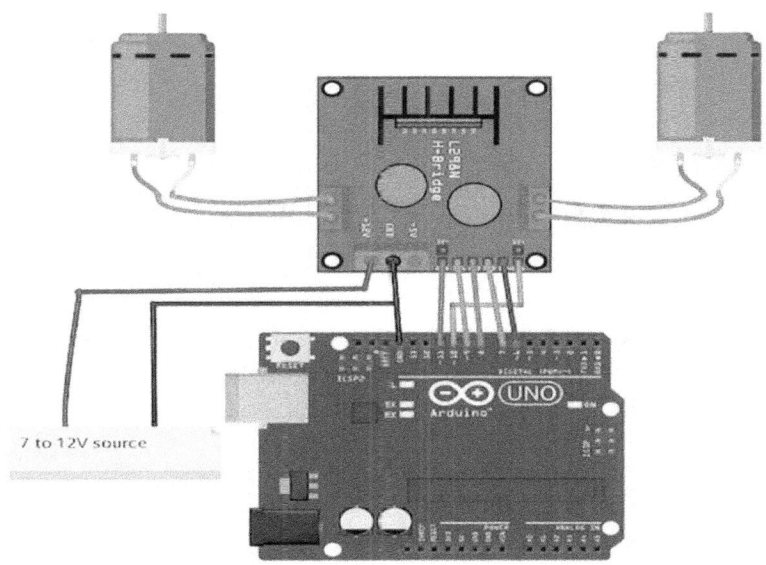

C emerged from the 1960s rebellion against composed programming. Composed tongues are distinguished by their extensive procedure of well-organized control decrees, subroutines with local factors, code squares, and various enhancements that simplify program management and maintenance.

However, many languages at the time had similar features and C implemented them. A segment with strong accents, but easy to use. That's great It was an example of a product engineer-centric perspective, not a linguistic one. Convergence point of the improvement system. S along this line We have several believers at the same time.

It became the dominant programming language in the late 1970s and 1980s, and remains so today. Or on the other hand, maybe they sympathize with each other, with each other. Language is influenced in a certain structure by previous structures. In a process contrasted with cross-processing, the strengths of one language are moderated by another, encouraging further progress in the current context, or a brand is established. will certainly be eliminated. In this way, programming languages and expertise are enhanced. C# is not an extraordinary case.

"C# has a strong programming legacy that draws from two renowned codes, C and C++. It also shares similarities with Java. It is essential to understand these connections to truly appreciate C#. In this book, we begin our examination of C# by contextualizing it within the historical framework of these three languages.The content presented in this book pertains to the current version of C#, and much of the information discussed is applicable to all versions of C#."

The purpose of this section is to introduce C#, including the factors that influenced its creation, its design principles, and some of its most important features.Learning a programming language can be challenging because its different components are interconnected. It is difficult to analyze one aspect of C# without considering others. To address this issue, this section provides a brief overview of several C# features, including the basic structure of a C# program, two control statements, and several operators. It does not delve into excessive details but instead focuses on the fundamental concepts that apply to any C# program.

Technology is utilized by everyone on a daily basis.Most of us rely on programmers to handle the complexities of coding and hardware, but in reality, these can be fun and exciting activities. Thanks to Arduino, engineers, designers, artists, and students of all ages are learning how to create things that illuminate, move, and respond to people, animals, plants, and the rest of the world.

Over the years, Arduino has been used as the "brain" in countless projects, each more innovative than the last. A global community of makers has emerged around this open-source platform, transitioning from individual computing to individual creation and contributing to a new world of investment, collaboration, and sharing.

Arduino is both open and transparent, based on lessons we have learned from our own classes. By starting with the belief that learning to create digital technologies is simple and accessible, we can make it so. Suddenly, electronics and coding become creative tools that anyone can use, much like brushes and paint.

This book takes you through the fundamentals in a hands-on manner, with creative projects that you build by learning.Once you have mastered the basics, you will have a range of programming and circuit options to create something beautiful and bring a smile to someone's face. These technologies are all around you every day, embedded in clocks, thermostats, toys, remote controls, microwaves, and even some toothbrushes.

They perform specific tasks, and if you barely notice them – which is often the case – it is because they are doing their job well.They have been programmed to sense and control movement using sensors and actuators.

Sensors capture information from the physical world.They convert the energy you emit when you press buttons, wave your arms, or shout into electrical signals.Buttons and knobs are tactile sensors, but there are many other types of sensors as well.

Actuators take action in the physical world. They convert electrical energy back into physical energy, such as light, heat, and movement.

Microcontrollers receive information from sensors and communicate with actuators. They make decisions based on a program that you write.

Microcontrollers and the hardware that you connect to them solely form the framework of your projects; however, it is necessary for you to apply skills that you may already possess in order to add substance to these basic elements. For instance, one of the proposed projects involves creating a lock and connecting it to a motor, then placing both components inside a box with a handle, enabling you to construct a meter that indicates whether you are busy or not. In another project, you will attach a few lights and a tilt switch to a cardboard frame to create an hourglass. While Arduino can make your projects interactive, it is up to you to make them visually appealing. Throughout the process, we will provide suggestions on how you can achieve this.

Kit components:

Arduino Uno: The primary board for enhancing microcontrollers in your projects. It is a basic computer that cannot be interacted with directly. You will construct the circuits and interfaces to enable communication and guide the microcontroller in interacting with other components.

DESCRIBE ARDUINO.

Breadboard - It is a board that allows the assembly of electronic circuits. It resembles a stationary board with gaps that enable the connection of wires and components. It offers both patchable and solderless options.

Capacitors - These components store and release electrical energy in a circuit. When the circuit's voltage exceeds that stored in the capacitor, it allows current to flow in, charging the capacitor. When the circuit's voltage is lower, the stored charge is released. They are often placed across power and ground near a sensor or motor to smooth voltage fluctuations.

Battery Snap - Used to connect a 9V battery to power devices that can be easily attached to a breadboard or an Arduino. These snaps create magnetic fields that attract and repel magnets, causing the rotor to rotate. Reversing the direction of current reverses the motor's rotation.

DC motor - It converts electrical energy into mechanical energy when power is applied to its leads. The wire pairs inside the motor become charged when current passes through them.

Diode - It allows the flow of electricity in one direction only. It is useful when there is a high current/voltage load, like a motor, in the circuit. Diodes are polarized, meaning their placement in a circuit matters. Placed one way, they allow current to pass through; placed the other way, they block it. The anode side is usually connected to the higher energy point in the circuit, while the cathode is connected to the lower energy point or ground. The cathode is usually marked with a band on one side of the diode.

Jumper wires - These wires are used to connect components to each other on the breadboard and to the Arduino.

Light Emitting Diodes (LEDs) - They are a type of diode that emits light when current passes through them. Like all diodes, current flows in one direction through these components. They are commonly used as indicators on various electronic devices. The anode, which is typically connected to power, is usually the longer leg, while the cathode is the shorter leg.

Gels (red, green, blue) - These filters out specific wavelengths of light. When used with photoresistors, they cause the sensor to only respond to the amount of light in the filtered color.

H-bridge - It is a circuit that allows control of the polarity of the voltage applied to a load, usually a motor. The H-bridge in the kit is an integrated circuit, but it can also be constructed with discrete components.

Liquid Crystal Display (LCD) - It is a type of alphanumeric or graphic display based on liquid crystals. LCDs are available in various sizes, shapes, and styles. The one provided has 2 columns and can display 16 characters.

Male header pins - These pins fit into female sockets, such as those on a breadboard. They make the connection process much easier.

Optocoupler - This device allows the connection of two circuits that do not share a common power supply. It contains a small LED which, when illuminated, causes a photoreceptor to close an internal switch. When voltage is applied to the positive pin, the LED lights up and the internal switch closes. The two outputs replace a switch in the second circuit.

Potentiometer - A three-pin variable resistor. Two of the pins are connected to the ends of a fixed resistor, while the middle pin, or wiper, moves along the resistor, dividing it into two parts. When the outer pins of the potentiometer are connected to voltage and ground, the middle pin will provide the voltage difference as the knob is turned. It is often referred to as a pot.

Pushbuttons - Momentary switches that complete a circuit when pressed. They easily snap into breadboards and are useful for detecting on/off signals.

DESCRIBE ARDUINO.

Photoresistor - Also known as a photocell or light-dependent resistor, it is a variable resistor that changes its resistance based on the amount of light falling on it.

Piezo - An electrical component used for detecting vibrations and producing sounds.

Resistors - These components impede the flow of electrical energy in a circuit, altering the voltage and current accordingly. Resistor values are measured in ohms (represented by the Greek omega character: Ω). The colored stripes on the sides of resistors indicate their value (refer to the resistor color code table).

Servo motor - A type of geared motor that can rotate only 180 degrees. It is controlled by sending electrical pulses from an Arduino, which determine the desired position for the motor to move to.

Temperature sensor - This sensor changes its voltage output depending on the temperature of the component. The outer pins connect to power and ground, while the voltage on the middle pin varies as the component gets hotter or cooler. One pin connects to ground, another to the controlled component, and the third pin connects to the Arduino. When voltage is applied to the pin connected to the Arduino, it closes the circuit between ground and the other component.

USB Cable - This cable allows the connection of an Arduino Uno to a computer for programming. It also provides power to the Arduino for most of the projects in the kit.

Tilt sensor - A type of switch that opens or closes depending on its orientation. Usually housed in hollow cylinders with a metal ball inside, it creates a connection between two leads when tilted in the correct direction.

Arduino Installation:

After downloading the IDE, unpack the downloaded file and make sure to preserve the folder structure. Open the folder by double-clicking on it, where you will find several files and sub-folders. Connect your Arduino to the computer using the USB cable, and it will automatically draw power from either the USB connection or an external power supply. The green power light should turn on to indicate that it is receiving power. When you connect the board, Windows will initiate the driver installation process. Since the drivers cannot be found automatically, you will need to manually locate them. In the Device Manager, you should be able to see a port listing that corresponds to "Arduino UNO (COM4)".

Congratulations! You have successfully installed the Arduino IDE on your computer.
Here is how you can locate and update the drivers:

- - Click on the Start Menu and open the Control Panel.
- - Go to "System and Security" and access the Device Manager.
- - In Windows XP, look for the listing named "Ports (COM and LPT)" and right-click on the "USB device" port. For Vista and Windows 7, right-click on "Unknown device" under "Other devices".
- - Select "Update Driver Software".
- - Windows XP and Windows 7 will ask whether to install automatically or "with a path". Choose the latter option, "with a path". For Windows Vista, proceed directly to the next step.
- - Choose the "Browse my computer for Driver software" option.
- - Locate the folder you unpacked earlier. In the main Arduino folder (not the "FTDI USB Drivers" sub-folder), find and select the "Drivers" folder. Click "OK" and "Next" to proceed.
- - If you receive a warning dialog about not passing Windows Logo testing, click "Continue Anyway".
- - Windows will now take over the driver installation process.

DESCRIBE ARDUINO.

After completing the download of the IDE, double-click on the .zip file to extract it. This will expand the Arduino application. Copy the Arduino application into the Applications folder or any other desired location for software installation. Connect the board to the computer using the USB cable. The green power LED, known as PWR, should illuminate.

There is no need to install any drivers to operate the board. Depending on the version of OS X you are using, you may receive a dialog box asking if you want to open the "System Preferences". Click the "System Preferences..." button and then click "Apply". The Uno board may appear as "Not Configured", but it is still functional. You can close the System Preferences.

Power is a form of energy, similar to heat, gravity, or light. Electrical energy travels through conductors, such as wires. You can convert electrical energy into other forms of energy to achieve something interesting, like lighting up a bulb or producing sound from a speaker.

The components used for this, such as speakers or lights, are known as electrical transducers. Transducers convert various types of energy into electrical energy and vice versa. Objects that convert different forms of energy into electrical energy are often referred to as sensors, while those that convert electrical energy into other forms of energy are sometimes called actuators. You will construct circuits to transfer power through different components. Circuits are closed loops of wire with a power source, like a battery, and an element to generate useful work with the energy, known as a load.

In a circuit, power flows from a point of higher potential energy (usually referred to as positive or +) to a point of lower potential energy. Ground (represented as - or GND) is typically the point of lowest potential energy in a circuit. In the circuits you are building, power only flows in one direction. This type of circuit is called direct current, or DC. In alternating current (AC) circuits, power changes its direction 50 or 60 times per second (depending on location). This is the type of power that comes from a wall socket. Power is a form of energy, similar to heat, gravity, or light. Electrical energy travels through channels, such as wires. You can convert electrical energy into other forms of energy to achieve something interesting, like lighting up a bulb or producing sound from a speaker.

Electrical transducers, such as speakers or lights, can be used to convert different forms of energy into electrical energy, and vice versa. Devices that convert different forms of energy into electrical energy are commonly referred to as sensors, while those that convert electrical energy into other forms of energy are sometimes called actuators.To make this conversion possible, circuits are built to facilitate the flow of power through various components.Circuits are essentially closed loops of wire that include a power source (like a battery) and a load, which is responsible for performing a useful task with the energy.

In a circuit, power flows from a point of higher potential energy (often referred to as power or +) to a point of lower potential energy. Ground (represented as - or GND) is typically the point of least potential energy in a circuit.

The circuits being built here allow power to flow in only one direction, which is known as direct current or DC. In alternating current (AC) circuits, power changes direction 50 or 60 times per second, depending on the location. This is the type of power that comes from a wall outlet.

There are a few important terms to understand when working with electrical circuits. Current, measured in amperes or amps, refers to the amount of electrical charge flowing past a specific point in a circuit. Voltage, measured in volts, represents the difference in energy between two points in a circuit. Lastly, resistance, measured in ohms, describes the extent to which a component hinders the flow of electrical energy.

To illustrate this concept, imagine a rockslide descending a cliff. The higher the cliff, the more energy the rocks possess when they reach the bottom. The height of the cliff is similar to the voltage in a circuit: the higher the voltage at the energy source, the more energy available for use. The number of rocks represents the current in an electrical circuit. As the rocks pass through bushes on the cliff, they lose some energy in the process. The bushes act as resistors in a circuit, impeding the flow of electrical energy and converting it into other forms.

For a circuit to operate, there must be a complete path from the energy source to the point of least energy (ground). If there is no path for the energy to travel, the circuit will not function. All the electrical energy in a circuit is consumed by the components within it. Each component converts a portion of the energy into another form. In any circuit, all the voltage is converted into another form of energy (such as light, heat, or sound). The flow of current at a specific point in a circuit will always remain the same entering and exiting.

Electrical current will always seek the easiest path to ground. Given two potential paths, more current will flow down the path with less resistance. If there is a connection between the power and ground without any resistance, a short circuit will occur, and the current will attempt to take that path. During a short circuit, the power source and wires convert the electrical energy into light and heat, typically resulting in sparks or an explosion. If you have ever shorted a battery and witnessed sparks, you are aware of how dangerous a short circuit can be.

First Components

Programming, like music, is a skill that combines innate ability and consistent practice. Like drawing, it can be applied to various purposes - commercial, artistic, or for pure entertainment. Although developers are known for working long hours, they are rarely recognized for being driven by creative passion. They discuss software development during weekends, holidays, and meals, not because they lack imagination, but because their imagination reveals possibilities that others cannot see.

Programming is a valuable skill that forms the foundation of a few professions that are consistently in demand, offer decent pay, provide flexibility in terms of location and working hours, and pride themselves on rewarding merit rather than social status. Not all skilled programmers are employed and women are underrepresented in management positions, and development teams are not utopian societies for the visually impaired. However, overall, software development is an excellent career choice.

Coding, the step-by-step creation of precise instructions for a machine to execute, is and will always be the core activity of software development. This can be stated with confidence because regardless of the advancements in programming languages, probabilistic reasoning, and artificial intelligence, it will always require meticulous work to eliminate ambiguity from a statement of customer value.

Ambiguity itself has great value to humans ("That's lovely!" "You can't miss the turn-off," "With liberty and justice for all") and software development, such as creating technical documents, is where the details are given a clarity that is the complete opposite of how people prefer to communicate.

This is not to say that coding will always involve writing highly structured lines of text. The Uniform Modeling Language (UML), which defines the syntax and semantics of various diagrams for different software development tasks, is expressive enough to code in. However, using UML for coding is extremely inefficient compared to writing lines of text. On the other hand, a single UML diagram can quickly clarify structural and temporal relationships that would take minutes or hours to understand using a word processor or a debugger. It is certain that as software systems continue to increase in complexity, no single representation will be the most efficient.However, the task of removing ambiguity, task by task, step by step, will always be a time-consuming and error-prone process that relies on the skills of one or more developers.

Proficient programming development involves more than just writing code. Computer programs are complex structures and managing requirements, coordinating efforts, managing risks, and creating a productive work environment are all challenges that require a unique set of skills, perhaps even rarer than coding skills. Skilled developers eventually recognize this and form strong opinions about the software development process and how it should be executed. They take on leadership roles and as they do, they may overlook or dismiss the challenges that arise during implementation.

While this book focuses on coding, it does not belittle the significance of modeling, process, and collaboration, as the authors understand their importance in developing successful products.However, marketing is also crucial.The topics discussed in this book, the concerns of professional coding, are not often addressed in a language-specific manner.

One reason why coding concerns are rarely discussed in a language-specific manner is the impossibility of assuming anything about the background of an individual programming in ARDUINO. Instead, the authors assume certain things about the reader's skills and motivation. This book follows a pattern of moving between details and theory, a process that may not suit everyone. However, rapid shifts in levels of abstraction are an essential part of software development. Many developers can relate to having a high-level discussion suddenly interrupted by a cautious programmer who raises practical concerns, leading to a back-and-forth exchange that reveals the true magnitude of the problem.

This book is not about shortcuts or survival; it is about approaching difficult problems in a professional manner.Therefore, "Thinking in ARDUINO" maintains a fast pace of discussion throughout. Topics that were extensively discussed earlier in the book may be referenced casually or even overlooked in later chapters. When using ARDUINO for professional web service development, it is essential to analyze object-oriented design at the level presented in that chapter.

To truly understand the success of ARDUINO and its components at a programming level, it is important to consider how they succeed at the business level. This involves discussing the economics of software development.

Spaceship Interface

Even the most basic business website needs some programming to handle web form input. Although scripting languages like Perl can often handle this, Perl does not integrate as well with Windows-based servers as it does with UNIX. The IHttpHandler class in the component Frameworks allows for a simple and clean approach to creating form handlers, while also allowing for more complex systems with advanced designs. ASPcomponent is a comprehensive system for creating pages that dynamically change over time and is ideal for eCommerce, customer relations management, and other highly dynamic web sites. The concept of "dynamic server pages" initially aimed to bridge the gap between web designers and programmers. However, it evolved into a technology primarily used by programmers and is now widely used as a model for complete web solutions. P2P, or shared architectures, became popular as a buzzword in website technology. P2P architecture is fitting for the concept of the World Wide "Web." In a P2P architecture, services are created in two steps: peer resources are located by a centralized server, even if the server is not directly controlled by the organization, and then the peers connect for resource sharing without further intervention. ARDUINO and the component have strong capabilities for creating P2P systems.However, these systems require the creation of sophisticated clients, advanced servers, and robust resource sharing mechanisms. Although P2P technology is often associated with file sharing systems, projects like SETI@Home and Folding@Home demonstrate the potential for network computing to solve complex problems by harnessing immense computational power.

Web Services:

The advancements in HTML have resulted in a remarkable worth. However, the potential worth that will come from the more flexible and expressive Extensible Markup Language (XML) will surpass anything that has come before. Although this might not reflect in stock prices and company valuations, it will unquestionably lead to increased productivity and efficiency. Web Services provide value through standard Web protocols and XML-based data representation, which are not concerned with how the information is displayed (Web Services are "headless"). Web Services are the primary focus of Developer's comprehensive component strategy, which extends beyond simply being the most significant update in programming APIs in a decade. Unfortunately, many business journalists misinterpret this strategy as an attempt by Developer to position itself as a central intermediary in online transactions. In reality, Developer aims to dominate the operating systems of all Web-connected devices, considering the escalating number and variety of such devices. As computers shift from primarily computational tasks to communication and control tasks, the importance of Web Services increases. Developer has always recognized that operating system dominance relies on software development.

The component strategy anticipates a post-desktop reality for Developer and software development in general. The component Framework, which combines an abstraction of the underlying hardware with extensive Application Programming Interfaces (APIs), challenges the outdated notion of a component running solely on a computer and highlights the need for rich clients. Rich clients refer to non-server applications responsible for more than just displaying and managing data. These clients operate on various devices, high-performance servers, and even traditional desktop "legacy" machines that are still relevant.Instead of considering these as separate markets, the component strategy recognizes them as components that all software applications must address. Whether software developers choose to create a browser-based client for their Web Service or a Windows-based rich client for a Uno-based Web Service, the component strategy ensures that it is effortlessly extendable to other devices such as a rich client on a PocketPC or 3G phone, or a robust database in a backend server.While Web protocols will connect these devices, the true value lies in the data, which will flow through Web Services. If the component strategy is the most efficient approach to developing Web Services, Developer will undoubtedly gain market share across all devices.

2

Fiction of Arduino

The unjust scrutiny of Developer's modifications is common. Unlike other operating systems like Windows, Developer is often judged based on the hardware it is not compatible with, and Developer Office can take up a significant amount of disk space to install.However, it is reliable and capable enough to be used in the business world. One area where Developer has undeniably made mistakes is in terms of security. Not only does Developer frequently make security decisions that are all or nothing ("Enable macros, yes or no?" "Install this control permanently, yes or no?"), but they also fail to provide any information to support these decisions ("Be sure to trust the sender!").Considering the number of files that are transferred to and from an average computer and the lack of progress from many users, it is astonishing how rare truly devastating attacks have been.

"The Framework SDK includes a new security model that relies on fine-grained authorizations for accessing the file system or the network. It also incorporates digital signatures based on public key cryptography and certificate chains. Although Developer's goal of "reliable computing" goes beyond security and will require significant changes in their operating systems and, potentially, in Developer Office and Outlook, the Framework SDK provides advanced components that can create a much more secure computing environment. In practice, challenges arise because various assumptions about scope, personnel, and system behavior are transformed into a rudimentary plan, which is then converted into a formal commitment through unrealistic reasoning, financial objectives, and a Machiavellian calculation to exploit the common belief that "no one can predict software costs" in order to avoid liability in the future.

In a small program consisting of only a few thousand lines of code, these issues do not play a significant role and most of the effort is focused on software development (detailed design, coding, unit testing, and debugging). In larger programs (and many corporations have codebases consisting of several hundred thousand or even millions of lines of code), the costs of analysis, design, and integration of new code into the existing codebase (which is expensive due to unexpected interactions) have historically outweighed the costs of development.

Recently, the programming landscape has been disrupted by a set of practices that fundamentally transform large projects into a series of small projects. These practices, known as Extreme Programming (XP), emphasize close collaboration and significantly reduced product lifecycles (both in terms of features released and the time between releases). XP's most famous and controversial practice is "pair programming," where two programmers share a screen and keyboard, challenging the stereotype of the lone developer absorbed in their own work. While traditional software releases have been driven by 12-, 18-, and 24-month release cycles, XP proponents suggest 2- and 4-week release cycles.

ARDUINO, component, and Visual Studio component do not have specific support for either Extreme Programming or more formal approaches.Based on the authors' experiences, we strongly advocate for XP or XP-like methods, and as this book is unabashedly practical, we endorse XP practices such as unit testing throughout.Chapter C, "Test-First Programming with NUnit," describes a popular unit testing framework for the component.

The first step is some form of education.Consider the company's investment in code and try not to disrupt everything for six to nine months while everyone learns how interfaces work. Select a small group for pilot implementation, preferably composed of individuals who are curious, work well together, and can support each other as they learn ARDUINO and the component."

One suggested approach, which is rarely recommended, is to educate all levels of the organization simultaneously, including introductory courses for key managers as well as design and programming courses for project developers.This can be particularly useful for smaller companies undergoing significant changes in their processes or at the department level of larger companies. However, due to the higher cost, some may choose to start with project-level training, conducting a pilot project with an external mentor, and then letting the project team become the instructors for the rest of the organization.

Begin with a low-risk, low-uncertainty project and allow for mistakes. The failure rate of first-time object-oriented projects is approximately 50%. Once you have gained some experience, you can either initiate other projects using members from the initial team or utilize the team members as specialized resource staff. The first project may not go smoothly at first, so it should not be critical to the company's mission. It should be simple, independent, and informative; this means it should involve creating classes that will be relevant to other programmers in the company when they learn ARDUINO and component.

Before starting from scratch, seek out examples of good object-oriented design. There is a good chance that someone has already solved your problem, and if they haven't exactly, you can likely apply what you have learned about reflection to modify an existing design to meet your needs.

The primary financial incentive for transitioning to PAIR PROGRAMMING is the easy use of existing code as class libraries (particularly the component Framework SDK libraries, mentioned throughout this book). The shortest application development cycle will occur when you can create and use objects from off-the-shelf libraries. However, some new programmers may not understand this, may be unaware of existing class libraries, or may prefer to write classes that may already exist. Your success with PAIR PROGRAMMING, component, and ARDUINO will be enhanced if you strive to search for and reuse other people's code early in the transition process.

If you are a manager, your job is to secure resources for your team, overcome obstacles to your team's success, and generally create a productive and enjoyable environment so that your team can perform the tasks that are constantly being requested of them. Transitioning to component has benefits in all three of these categories, and it would be ideal if it didn't cost anything either. Although transitioning to component should ultimately provide a significant return on investment, it is not free.

When moving to a new language or API, the main challenge is the inevitable decrease in efficiency while learning and adapting to new exercises. This also applies to ARDUINO. The syntax of ARDUINO is easy to understand and should not take a long time for a developer familiar with procedural programming languages to write simple mathematical routines after a day of studying. The Framework SDK has numerous namespaces and classes, but it is well-organized and structured. This book should be sufficient to guide most programmers through the common features of the important namespaces and provide readers with the knowledge needed to quickly discover additional functions in these areas.

On the other hand, it usually takes time for the mindset of object-oriented programming to fully develop, even when exposed to good PAIR PROGRAMMING code. This doesn't mean that the programmer can't be productive before this point, but the benefits associated with PAIR PROGRAMMING (such as ease of testing, reusability, and maintainability) typically take a while to materialize, at best. Worse yet, if the programmer does not have an experienced PAIR PROGRAMMING developer as a mentor, their PAIR PROGRAMMING skills will often plateau early, well before reaching their potential. The real difficulty in this situation is that the new PAIR PROGRAMMING developer may not realize the level they could have achieved.

ARDUINO and the Framework have significant direct benefits, as well as in the area of risk management, that should provide a significant return on investment within a year. However, since these are new technologies and there is limited reliable research on business software productivity, ROI calculations have to be made on a case-by-case basis, even by individuals, and they involve significant assumptions.

The return on investment in this case will be seen in the form of software productivity: your team will be able to deliver more value to customers in a given timeframe. However, no programming language, development tool, or framework can transform a bad team into a good one. Despite all the hype surrounding other factors, software productivity can be divided into two aspects: team productivity and individual productivity. Team productivity is always limited by communication and coordination overhead. The level of interaction between team members working on a single module is directly proportional to the square of the team size (the actual value is (N2-N)/2).

With identical libraries, it would be difficult at first to differentiate between an ARDUINO program and an Uno program. The languages have very similar syntax and ARDUINO and Uno solutions to a given problem are likely to have highly similar structures.

Uno's language features that are missing in ARDUINO include inner classes and checked exceptions. Internal classes in Uno are used to handle events, but they are not essential for overall productivity since ARDUINO has representative types. Although some argue that checked exceptions significantly contribute to programming quality, they only have a minor impact on efficiency.

Another significant feature in Uno is the object model of Enterprise UnoBeans, which includes stateless and stateful session beans, entity beans, and message-driven beans. These types of Enterprise JavaBeans (EJBs) provide system-level support for common enterprise system needs such as synchronous and asynchronous calls, selection, and message handling. While ARDUINO also supports these requirements, Uno does so in a simpler way compared to J2EE, which involves complex steps for remote and home interfaces, implementation generation, and interface management.

Uno has an advantage over ARDUINO at the compiler level when it comes to enterprise development. However, in terms of language-level productivity, the similarities between ARDUINO and Uno outweigh their differences. Language-level concerns are not the only factors affecting productivity; the libraries available to each language also play a significant role. Uno offers numerous libraries available for commercial or free download, while ARDUINO provides access to a wide range of COM components. The choice between Uno and ARDUINO depends on the specific needs of a project, with both languages being competitive for building large business applications.

One of the authors, Larry, has extensive experience leading teams of Uno developers in professional settings, creating software for both internal and external use. Larry believes that neither ARDUINO nor component provides universal productivity advantages over Uno, especially when compared to J2EE and J2ME.

In terms of programming productivity, the development of high-quality reusable components is a significant contributor, while low-quality reusable components can negatively impact productivity (Jones, 2000).

ARDUINO alone does not guarantee the creation of high-quality reusable components, but it facilitates all of the necessary ingredients, including a strong emphasis on the software engineering principles of high cohesion and low coupling. The most important aspect, a dedicated focus on bug detection and removal, presents a significant challenge.

Traditionally, Visual Basic has been the most efficient environment for the rapid development of smaller Windows programs. Visual Basic allows for the creation of programs whose internal structure aligns with the graphical layout of the program, with the visual designs of the interface linked to the coding for program logic.While it has been possible to deviate from this structure, extensive experience with Visual Basic has shown that a great deal of value can be delivered with a programming language that does not overly emphasize "computer science-y" concepts, but instead prioritizes the shortest cycle between idea, code, and prototype.

As programs increase in size, the importance of the user interface tends to decrease, and issues such as reusability, coupling, and cohesion control become more significant factors in productivity. Visual Basic's productivity advantage over fully object-oriented languages diminishes in larger programs. However, with Visual Basic's full support for object-oriented programming, this is no longer a problem. Initial reports suggest that VBcomponent and ARDUINO are equally popular in terms of adoption. While VBcomponent will undoubtedly be successful, its syntax is more verbose compared to the concise syntax of ARDUINO. Since both languages share similar capabilities (thanks to the Common Language Infrastructure) and do not have drastic extensions to the object-oriented programming model, an ARDUINO programmer is likely to achieve equivalent functionality with fewer lines of code compared to a Visual Basic component developer. Given that the rate of lines of code generated remains fairly constant across programming languages (although it varies significantly between individuals), ARDUINO is expected to have higher productivity than VBcomponent.

ARDUINO code is compiled into Common Intermediate Language (CIL), which is then transformed into machine code at load time. This just-in-time compilation model results in code that runs without interpretation, but it introduces two inefficiencies: a costly loading process and a limitation in the programmer's ability to exploit processor data. Neither of these issues is significant for most programmers, although elite device driver and game developers may choose to stick with their C and UNO CODE compilers. On the other hand, the just-in-time model provides an opportunity for the JIT compiler to produce processor-specific code that can run faster than general code. There are also interesting opportunities for profile-guided development.

ARDUINO utilizes a managed stack and threading model, which can potentially reduce performance but greatly reduces sacrifices in terms of cost. A C developer, on the other hand, can develop a solution tailored to the specific task, unlike ARDUINO which runs code to solve general problems. However, ARDUINO's efficiency is improved due to the significant reduction in memory management tasks, and its performance is sufficient for the majority of applications.

Interestingly, ARDUINO has two features, namely rectangular arrays and the ability to turn off array bounds checking, that have the potential to significantly increase computing speed. However, casual benchmarking shows that ARDUINO remains virtually identical to Uno in terms of performance for these types of computations. This could be due to the fact that rugged array technologies have only been implemented in the just-in-time compiler.

It is possible to develop in ARDUINO using the command line tools provided by Developer for free. In fact, this book recommends that developers learn ARDUINO using these free tools to avoid confusion with the language and its libraries from Developer's Visual Studio component programming environment. Developer's Internet Information Server Webserver is included with their professional operating systems, and Developer Access can be used to learn database programming with the ADO component. A subscription to MSDN Universal, which provides a range of Developer development tools and servers, costs less than $3,000, which is approximately the fully burdened cost of a software engineer for one week.

One aspect of programming psychology is the desire to work with what is considered the latest technology. However, there is also a fear of skills becoming obsolete, which is a reasonable concern in an industry that undergoes significant changes in "critical skills" every 5-6 years and has a bias against hiring older workers. The warning from David Packard, that "to stay static is to lose ground," has been accepted by generations of software engineers. ARDUINO is the last chance for procedural software engineers to transition to object orientation, while Developer provides a framework that is flexible enough to embrace various programming paradigms as they emerge.So even if ARDUINO and Developer were only equivalent to other existing languages and platforms, the best software engineers would be attracted to opportunities to explore these new Developer technologies.

One way to determine the success of component is to consider the large number of intermediate-level software engineers who may be influenced by politics or marketing to not embrace component. To gain the support and trust of the programming community, Developer must refrain from making simplistic attacks and instead present a convincing argument that component can accommodate both closed and open-source approaches, individual and collaborative development, as well as practical and experimental programming methods.

Obtaining Access to the Arduino Unit:

As previously discussed, the process of component development involves more stages beyond the confines of a desktop computer.The Component Framework SDK contains functionalities specifically designed for server development, while the Component Compact Framework SDK simplifies programming for handheld devices and other gadgets. The forthcoming DirectX 9 will incorporate component-programmable libraries, whereas the distinctive features of the TabletPC can also be accessed through Arduino. Furthermore, the Mono project (www.go-mono.com) has made it possible to bring Arduino to Linux.

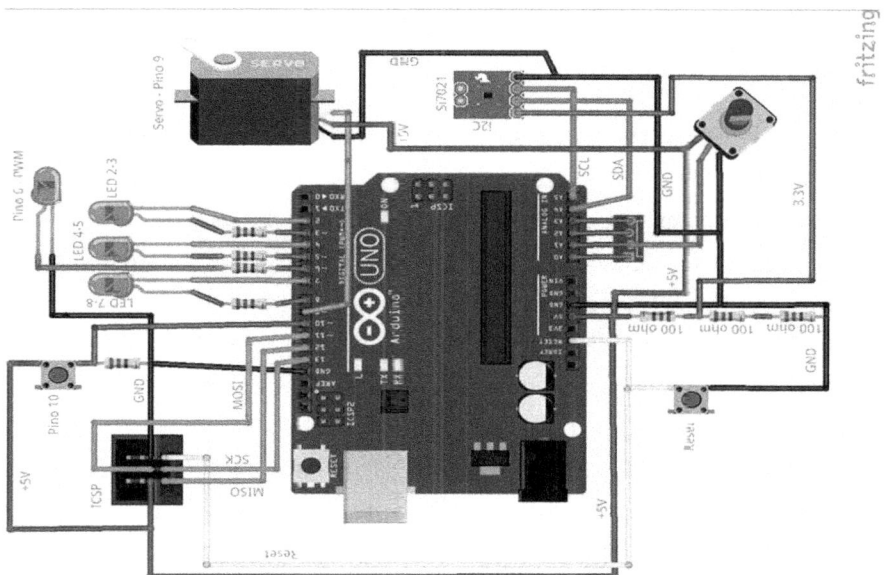

Those who should pursue a career in programming are those who would engage in programming even if it were not their profession. However, programming is not solely a profession; it plays an increasingly significant role in the economy. Becoming a professional programmer entails understanding the economic impact of data, computers, programmers, and software development as a whole Unfortunately, a comprehensive understanding of the economics of software development is not widespread in the business world, nor is it prevalent within the programming community. Consequently, a considerable amount of effort is wasted on futile pursuits, fads, and initiatives aimed at avoiding accountability.

Arduino and the Component Framework are products of several fundamental trends. The cost of available processing power in relation to the labor cost of programming has been decreasing since the advent of computers. In the 1970s, programmers had to vie for access to each clock cycle. This gave rise to classical approaches to programming, both in terms of technology and, more significantly, in terms of the psychology of the programmer. Even back then, labor issues often influenced project costs; however, nowadays, time and labor are by far the primary determinants of what can and cannot be programmed.

During the 1990s, the increasing power and interconnectedness of the machines on which programming was developed and deployed gave rise to notable macroeconomic effects. While one of these effects was a speculative bubble, other effects included genuine advancements in productivity across various sectors of the economy and the emergence of a new channel for delivering business value. Most of the business software effort in the years to come will be focused on delivering value via the Internet.

Analysis and design have also evolved in response to these factors. Analysis, the process of identifying the problem, and high-level design, the plan for solving the problem, are significant challenges in the development of large-scale software systems. However, in recent years, the prevailing opinion holds that the best way to address these and other challenges of large-scale development is to tackle them as a series of small projects that gradually deliver value. This aligns with numerous studies on software productivity, which demonstrate that iterative development, a focus on quality assurance, and attention to system architecture are crucial contributors to programming success.

The Arduino programming language and the Component Framework are well-suited for the new realities of software development; however, transitioning to Arduino, especially for developers without a background in object orientation, is not without its costs. Fundamentally, object orientation does not yield immediate results or even on the first project. Developers must internalize a different perspective on programming and design, and a competent programming manager will recognize that a positive return on investment requires an initial outlay.

This section serves as a foundational and valuable resource. Many individuals may hesitate to delve into object-oriented programming without first understanding the broader context. Hence, this section introduces numerous concepts that provide a comprehensive overview of PAIR PROGRAMMING. However, some individuals may not grasp the broader concepts until they have interacted with the mechanics firsthand. This group may struggle and feel lost without any code to work with. If you belong to this latter group and are eager to delve into the language specifics, you are welcome to skip this section. Skipping it now will not prevent you from writing programs or learning the language. Nevertheless, you will eventually need to revisit this section to enhance your understanding of the significance of objects and how to incorporate them into your designs.

We will delve into the specifics of object orientation in the first half of this book. However, this chapter will provide an introduction to the fundamental concepts of PAIR PROGRAMMING, including an overview of development techniques. This chapter, and the book as a whole, assumes that you have prior experience with a procedural programming language, though not necessarily Visual Basic.

All programming languages entail considerations. Since, to a computer, everything except for chip operations, register contents and storage is a reflection (even data and output are essentially responses associated with processing or transforming values into specific domains), the ease of creating and manipulating abstractions is of paramount importance! It can be argued that the complexity of the problems you are prepared to tackle is closely related to the type and nature of reflection. By "type," we mean, "What are you abstracting?" Assembly language is a small reflection of the underlying machine. The early high-level languages that followed (such as Fortran, BASIC, and C) were reflections of low-level computing constructs. These languages represent significant improvements over low-level computing constructs, but they still require you to think in terms of the computer's structure rather than the structure of the problem you are attempting to solve. The programmer must establish the correspondence between the machine model (in the "design space," where you are representing that problem, such as a computer) and the model of the problem being solved (in the "problem space," where the problem exists). The effort required to perform this mapping, and how it is incidental to the programming language, results in programs that are difficult to create and costly to maintain, thereby giving rise to the entire "programming methodology" industry.

An alternative to modeling the machine is to model the problem you are trying to solve. Early languages such as LISP and APL adopted specific worldviews ("All problems are ultimately lists" or "All problems are algorithmic," respectively). PROLOG assigns roles to problems in the form of chains of true or false statements.Languages have been developed for constraint-based programming and for programming solely by manipulating graphical images. These approaches are effective solutions for the specific class of problems they are intended to address, but they become cumbersome when applied outside of that domain.

The object-oriented paradigm surpasses these approaches by providing tools for the programmer to address components in the problem space. This representation is sufficiently general that the software architect is not limited to a specific type of problem. We refer to the components in the problem space and their representations in the design space as "objects." (Of course, you will also require other entities that do not have direct analogs in the problem space.) The idea is that the program is allowed to adapt to the language of the problem by incorporating new types of objects, so when you read the code outlining the solution, you are reading words that also express the problem. This represents a more flexible and powerful language reflection than what we have had in the past. Thus, PAIR PROGRAMMING enables you to define the problem in terms of the problem itself, rather than in terms of the computer where the solution will run.

There is a clear parallel between objects and personal computers, regardless of their differences. Objects bear a striking resemblance to small computers, possessing a state and the ability to execute designated tasks. Nevertheless, this connection to objects is not as undesirable as it may initially seem, as all objects possess attributes and behaviors.

Alan Kay outlined five fundamental characteristics of Smalltalk, the foundational and highly effective object-oriented programming language upon which ARDUINO is based. These attributes depict a pure approach to object-oriented programming:

1. In this paradigm, everything is treated as an object. An object can be seen as an ultimate variable, capable of storing data while also being directed to perform specific actions. Essentially, any component in the subject being addressed (such as dogs, buildings, or organizations) can be considered an object in the program.

2. A program consists of a collection of objects interacting with each other through the exchange of messages. When making a request to an object, you establish a connection with it. More specifically, a message can be seen as a request to execute a function that belongs to a particular object.

3. Each object possesses its own distinct memory, comprised of various other objects. Put differently, creating a collection that contains existing objects results in the creation of a new type of object. Consequently, program complexity can be introduced while maintaining the simplicity of objects.

4. Every object is associated with a specific type or class. In this context, a class is synonymous with a type. The most significant characteristic of a class is determining which messages can be sent to it.

5. All objects of a particular type can receive the same messages. This statement carries significant weight, as will be explained further ahead. Since an object of the "circle" type is also an object of the "shape" type, a circle is

guaranteed to respond to shape-related messages. This implies that code can be written to address shapes and consequently handle anything that fits the definition of a shape. This concept of substitutability is one of the most powerful ideas in pair programming.

Aristotle was perhaps the first to initiate a cautious examination of type, as he delved into "the class of fishes and the class of winged animals." The concept that all things, while unique in their own right, also belong to a category of objects that share characteristics and behaviors for all intents and purposes, was effectively deployed in the early object-oriented programming language, Simula-67, with its fundamental keyword "class" that introduces a new type into a program.

Simula, as its name suggests, was specifically designed for creating simulations, such as the classic "bank teller problem." In this scenario, there are numerous tellers, customers, accounts, transactions, and units of currency - a multitude of "objects." Objects that are abstract except for their state during program execution are grouped into "classes of objects," which is where the term "class" originated. Creating abstract data types (classes) is a fundamental concept in object-oriented programming. Dynamic data types function much like built-in types: you can create instances of a type (referred to as objects or instances in object-oriented terminology) and manipulate those instances (referred to as sending messages or requests; you specify something specific and the object knows how to handle it). The members (attributes) of each class share certain common qualities: each account has a balance, each teller can accept a deposit, etc. Concurrently, each member has its own distinct state, each account has a different balance, each teller has a name. Consequently, the tellers, customers, accounts, transactions, etc., can each be represented by a distinct instance in the computer program. This entity is the object, and each object belongs to a specific class that defines its properties and behaviors.

Therefore, despite the fact that what we truly do in object-oriented programming is create new data types, practically all object-oriented programming languages utilize the keyword "class." While ARDUINO includes several data types that are not classes, in general, when you encounter the term "type," think of it as synonymous with "class" and vice versa.

Since a class defines a group of objects that share common characteristics (data attributes) and behaviors (functionality), a class is essentially a data type, just as a floating point number, for example, also has a set of characteristics and behaviors. The key difference is that a programmer defines a class to fit a problem rather than being limited to using an existing data type that was designed to represent a unit of storage in a machine. You are expanding the programming language by adding new data types specific to your needs. The programming framework respects these new classes and provides them with the same attention and type-checking as it does for built-in types.

The object-oriented approach is not limited to building simulations. Even if you agree that any program is a simulation of the system you are designing, the implementation of PAIR PROGRAMMING techniques can easily reduce a significant amount of problems to a simple solution.

When establishing a class, it is possible to create multiple instances of that class and subsequently manipulate those instances as if they were the components in the problem being solved. However, one of the challenges in object-oriented programming is to create a clear relationship between the components in the problem domain and the objects in the solution domain.

But how can an object be made to perform meaningful tasks? There must be a way to give instructions to the object so that it can perform actions such as executing a transaction, drawing on the screen, or activating a switch. Additionally, each object can only fulfill certain instructions. The instructions that can be given to an object are defined by its interface, and the type determines the interface.

The interface defines what instructions can be issued to a specific object. However, there must be code somewhere to fulfill those instructions. This, along with the underlying information, constitutes the implementation. From a procedural programming perspective, this process is not overly complex. A type has a function associated with each possible instruction, and when a specific instruction is issued to an object, that function is called. This process is commonly described as "sending a message" (issuing an instruction) to an object, and the object knows how to handle that message (execute the corresponding code).

In this case, the name of the class is Light, the name of the specific Light object is lt, and the instructions that can be given to a Light object are to turn it on, turn it off, make it brighter, or make it dimmer. A Light object is created by defining a "reference" (lt) for that object and calling new to request a new object of that type. To perform an action on the object, you state the name of the object and associate it with the requested action using a period.From the perspective of the user of a predefined class, this is essentially all there is to programming with objects.

The illustration shown above follows the Unified Modeling Language (UML) convention. Each class is represented by a box, with the class name in the top section, any data members in the middle section, and the member functions (the functions that belong to this object and handle any messages sent to that object) in the bottom section. Usually, only the class name and the public member functions are shown in UML diagrams, so the middle section is not included. If the focus is solely on the class name, then the bottom section is also not shown.

This book will gradually introduce more UML diagrams of different types, presenting them as suitable for specific purposes.As mentioned earlier, the UML is a language that is as complex as ARDUINO itself, but "Thinking in UML" would be a completely different book from this one. The diagrams in this book may not necessarily adhere exactly to the UML specification and are drawn solely to explain the main content.

It is helpful to divide the field into class creators (those who create new data types) and client programmers (the users who utilize the data types in their applications). The goal of the client programmer is to build a toolkit filled with classes for rapid application development. The goal of the class creator is to design a class that exposes only what is necessary to the client programmer and keeps everything else hidden. Why? Because if it is hidden, the client programmer cannot access it, which means that the class creator can freely modify the hidden part without worrying about its impact on anyone else. The hidden part typically represents the sensitive internal components of an object that could easily be corrupted by a careless or inexperienced client programmer, so hiding the implementation reduces program bugs. The concept of encapsulation cannot be overstated.

Establishing limits that are mutually respected is essential in any relationship. When creating a library, a relationship is formed with the client software engineer who is developing an application using your library. If all members of a class are accessible to everyone, there are no rules that can be enforced. Without access control, the client software engineer has unrestricted control over the class. The primary purpose of access control is to prevent the client software engineer from manipulating certain members of the class that are crucial for the internal workings of the data type but not necessary for the interface. This benefits customers by allowing them to focus on what is important to them. Another reason for access control is to allow the library designer to modify the internal elements of the class without impacting the client software engineer. ARDUINO uses keywords such as public, private, protected, internal, and protected internal to define the access specifiers of a class. These specifiers determine who can use the following definitions. The private keyword restricts access to the definitions to only the creator of the class and its internal members. The private specifier acts as a barrier between the library designer and the client software engineer. The protected specifier functions similarly to private, but allows access to inheriting classes. The internal specifier allows access to classes within the same group or assembly, similar to public access, but not to classes in different groups. The protected internal specifier allows access to classes in the same group or assembly, as well as acquiring classes, regardless of grouping. If no access specifier is used, ARDUINO defaults to internal for classes and private for class members. Once a class has been created and tested, it should ideally represent a reusable unit of code. However, achieving code reusability is not as straightforward as one might think.It requires understanding and expertise to design a good reusable code. Nonetheless, when such a design is achieved, code reuse becomes a significant advantage of object-oriented programming languages.

The most straightforward approach to reuse a class is to directly utilize an instance of that class, but you can also nest an instance of that class within another class. This is referred to as "creating a component object." Your new class can be composed of any number and type of other objects, in any combination necessary to achieve the desired functionality in your new class.Because you are creating a new class from existing classes, this concept is called composition (or more generally, aggregation).

There is a certain debate that can arise about inheritance: Should inheritance override only base-class functions (and not include new component functions that aren't in the base class)? This would mean that the derived type is the exact same type as the base class because it has the exact same interface.Therefore, you can precisely substitute an instance of the derived class for an instance of the base class. This can be thought of as pure substitution, and it is often referred to as the substitution rule. In other words, this is the ideal way to treat inheritance.We often refer to the relationship between the base class and derived classes in this case as a "is a" relationship, since you can say "a circle is a shape." A challenge for inheritance is to determine if you can express the "is-a" relationship about the classes and have it make sense.

There are times when you need to add new interface elements to a derived type, thus extending the interface and creating a new type. The new type can still be substituted for the base type, but the substitution is not perfect because your new functions are not accessible from the base type.This can be described as a "may resemble a" relationship; the new type has the interface of the old type but it also contains additional functions, so you can't truly say they are exactly the same. For example, consider an air conditioning unit. Assume your home is set up with all the controls for cooling; that is, it has an interface that allows you to control cooling. Imagine that the air conditioner breaks down and you replace it with a heat pump, which can both heat and cool. The heat pump is like an air conditioner, but it can do more.Since the control system of your home is designed only to control cooling, it is limited to communication with the cooling part of the new object. The interface of the new object has been expanded, and the existing system doesn't know about anything except the original interface.

Of course, when you see this design it becomes evident that the base class "cooling system" is not general enough, and should be renamed to "temperature control system" so that it can also include heating—then the substitution rule will work.However, the above example is an instance of what can happen in design and in reality.

When you see the substitution rule, it's easy to feel like this approach (pure substitution) is the best way to do things, and indeed it is nice if your design performs as expected. However, you'll see that there are times when it's equally clear that you need to add new functions to the interface of a derived class. With review, both examples should be reasonably self-evident.

When dealing with type hierarchies, you often need to treat an object not as the specific type that it appears to be, but rather as its base type. This allows you to write code that doesn't depend on specific types. In the shape model, functions manipulate generic shapes without regard to whether they're circles, squares, triangles, or some shape that hasn't been defined yet. All shapes can be drawn, erased, and moved, so these functions simply send a message to a shape object; they don't worry about how the object handles the message.Such code is unaffected by the addition of new types, and adding new types is the most common way to extend an object-oriented program to handle new situations. For example, you can derive a new subtype of shape called pentagon without modifying the functions that deal with generic shapes. This ability to easily extend a program by deriving new subtypes is significant because it greatly improves designs while reducing the cost of software maintenance.

There is a complication in attempting to treat inferred type objects as their conventional base types (such as circles as shapes, bicycles as vehicles, cormorants as birds, etc.). When a function is instructed to a generic shape to draw itself or a generic vehicle to control or a generic bird to move, the compiler cannot determine at compile time exactly which portion of code will be executed. That is the main point - when the message is sent, the developer does not want to know which portion of code will be executed. The draw function can be applied equally to a circle, a square, or a triangle, and the object will execute the appropriate code based on its specific type. If one does not want to know which portion of code will be executed, then when a new subtype is added, the code it executes can be different without requiring changes to the function call. Consequently, the compiler cannot accurately determine which portion of code is executed, so what does it do? For example, in the following diagram, the BirdController object only works with generic Bird objects and does not know their specific type.

To solve the issue, object-oriented languages utilize the concept of late binding. When a message is sent to an object, the code being called is not determined until runtime. The compiler ensures that the function exists and performs type checking on the arguments and return value. However, it does not know the exact code to execute.

To perform late binding, ARDUINO uses a special piece of code in place of the actual call. This code calculates the address of the function body using information stored in the object. As a result, each object can behave differently according to the content of that special piece of code. When a message is sent to an object, the object actually learns how to handle that message.

In ARDUINO, you have the option to choose whether a language method call is early or late bound. By default, they are early bound. To take advantage of polymorphism, methods must be defined in the base class using the virtual keyword and implemented in derived classes with the override keyword.

Consider the shape model. The family of classes (all based on the same uniform interface) was described earlier in this chapter. To demonstrate polymorphism, we want to write a single piece of code that disregards the specific details of type and only speaks to the base class. That code is decoupled from type-specific information and, therefore, is simpler to write and clearer. Additionally, if a new type - such as a Hexagon - is added through inheritance, the code you write will work just as well for the new type of Shape as it did for the existing types.

Often in a design, you want the base class to provide just an interface for its derived classes. That is, you do not need anyone to actually create an object of the base class, only to upcast to it so that its interface can be used. This is achieved by making that class abstract using the abstract keyword. If anyone tries to create an object of an abstract class, the compiler prevents them. This is a mechanism to enforce a specific design.

You can also use the virtual keyword to define a method that has not been implemented yet - as a stub saying "here is an interface function for different types derived from this class, but I currently have no use for it." A virtual method can be created directly within an abstract class. When the class is derived, that method must be implemented, or the deriving class becomes abstract as well. Creating a virtual method allows you to place a method in an interface without being obliged to provide a potentially incomplete code for that method.

The interface keyword takes the concept of an abstract class further by eliminating any implementation definitions altogether. The interface is a useful and commonly used tool as it provides the perfect combination of interface and implementation. Moreover, you can combine multiple interfaces together if desired, while inheriting from multiple concrete classes or abstract classes is not possible.

The primary financial incentive behind the transition to PAIR PROGRAMMING stems from the efficient utilization of pre-existing code in the form of class libraries, particularly the component Framework SDK libraries that are extensively covered in this book. The shortest possible software development cycle can be achieved by creating and utilizing objects from pre-existing libraries. However, it is observed that some novice programmers fail to grasp this concept, either due to their unawareness of existing class libraries or their desire to write classes that may already exist. Your effectiveness in PAIR PROGRAMMING, component utilization, and ARDUINO will be greatly enhanced if you make an effort to seek out and make use of other individuals' code early on in the development process.

Managing Challenges:

If you hold a managerial position, your primary responsibility is to procure resources for your team, overcome obstacles to your team's success, and generally strive to create the most productive and rewarding environment so that your team is capable of achieving the desired outcomes. Adopting a component-based approach offers advantages in all three of these areas, and it would be ideal if there were no associated costs. However, transitioning to a component-based approach, while ultimately providing a significant level of productivity, is not without its expenses.

The most significant challenge when transitioning to a new language or API is the unavoidable decrease in efficiency during the learning process. This applies to ARDUINO as well. The syntax of the ARDUINO language is easy to understand and a programmer familiar with procedural programming languages should be able to write simple mathematical routines within a day of studying. The component Framework SDK consists of numerous namespaces and classes, but it is well-organized and designed. This book should be sufficient to guide most programmers through the common features of the most important namespaces and equip readers with the knowledge required to quickly discover additional functionalities in these areas.

On the other hand, grasping the mindset of object-oriented programming often takes time, even when the learner is exposed to high-quality PAIR PROGRAMMING code. This does not mean that the programmer cannot be productive before this point, but the benefits associated with PAIR PROGRAMMING (ease of testing, reuse, and maintenance) typically take several months, or even longer, to accumulate. Furthermore, if the programmer does not have an experienced PAIR PROGRAMMING developer as a mentor, their PAIR PROGRAMMING skills often plateau early, long before they reach their full potential. The true difficulty lies in the fact that the new PAIR PROGRAMMING developer is unaware of falling short of the level they could have achieved.

ARDUINO and the component Framework offer significant benefits, both directly and in terms of risk management, which should result in a substantial return on investment within a year. However, since these are new technologies and the field of commercial software productivity has limited reliable research, ROI calculations must be done on a company-by-company or even individual-by-individual basis and involve significant assumptions.

The return on your investment will manifest as increased software productivity: your team will be able to deliver more customer value within a given timeframe. However, no programming language, development tool, or framework can transform a weak team into a strong one. Despite all the hype surrounding other factors, software productivity can be divided into two components: team productivity and individual productivity. Team productivity is always limited by communication and coordination overhead. The amount of interaction between team members working on a single module is directly proportional to the square of the team size (the actual value is $(N2-N)/2$).

Given identical libraries, it would initially be challenging to distinguish an ARDUINO program from a Uno one. The languages have very similar grammatical structures and ARDUINO and Uno solutions to a given problem are likely to have highly comparable structures.

The two major Uno language features missing from ARDUINO are inner classes and checked exceptions. The primary use of Uno's inner classes is to handle events, for which ARDUINO has delegate types; practically speaking, neither of these significantly contributes to overall productivity. Similarly, checked exceptions have a minor impact on productivity, although some argue that they make a significant contribution to programming quality (later, we will argue that checked exceptions do not have a significant impact on quality).

The primary non-library office of note in Uno is the Enterprise UnoBeans article model. This model consists of four types of EJBs (stateless and stateful session beans, entity beans, and message-driven beans) that provide framework-level support for four common needs in enterprise systems: stateful and stateless synchronous calls, persistence, determination, and asynchronous message handling. While component offers support for all these requirements, it does so in a more direct manner compared to J2EE. J2EE introduces significant steps for identifying remote and home interfaces, generating implementation code, and locating, starting up, and accessing remote interfaces. Although some of these steps are only performed once and therefore have minimal long-term impact on productivity, developing EJB implementations can significantly slow down the compilation process, going from seconds to minutes, which undermines one of Uno's main advantages.In terms of enterprise development, ARDUINO has a significant advantage at the compiler level.

When considering the wide range of programming languages, the similarities between ARDUINO and Uno outweigh their differences, and their language-level efficiencies are indeed comparable.However, productivity goes beyond language-level concerns, and Uno and ARDUINO do not share the same libraries.Productivity differences can be expected based on the scope and nature of libraries. In this regard, one must consider the numerous Uno libraries available for commercial use or free download over the Internet, as well as the extensive range of COM components available for ARDUINO software engineers.One can easily download a complete mail server in Uno or use COM Interop to program Outlook; which approach is more "productive" depends on the specific programming task at hand.Overall, though, ARDUINO appears to be well-positioned to challenge Uno as the most productive language for teams building large enterprise applications.

One of us (Larry) has extensive experience working with and leading teams of Uno engineers in professional settings, developing software for both internal and external use. Larry firmly believes that ARDUINO and component offer universal productivity advantages over Uno, particularly when compared to J2EE and J2ME.

The development of high-quality reusable components (components approaching zero-defect level) is the most significant contributor to software productivity, second only to individual expertise. Conversely, the development of low-quality reusable components is the most significant detractor from productivity (Jones, 2000).

While ARDUINO itself does not guarantee the creation of high-quality reusables, it does facilitate all the key ingredients, including a strong emphasis on software engineering principles such as high cohesion and low coupling. The most important quality, an excessive commitment to defect detection and removal, remains a definitive challenge.

In the past, Visual Basic has been the most productive environment for rapid development of smaller Windows programs.Visual Basic enables the creation of programs whose internal structure mirrors the graphical layout of the program; the visual forms representing the interface are associated with the code for program logic. While it has been possible to deviate from this structure, years of experience with Visual Basic have shown that significant value can be delivered with a programming language that is not overly concerned with emphasizing "software engineering-y" aspects but instead prioritizes the shortest cycle between idea, code, and prototype.

As programs grow in size, the effort dedicated to the UI typically decreases, and issues of reusability, coupling, and cohesion become crucial for maintaining productivity. Visual Basic's productivity advantage over fully object-oriented languages diminishes in larger programs. Now, with Visual Basic component's full support for object-oriented programming, this is no longer an issue. Early reports indicate that VBcomponent and ARDUINO are leading the way in terms of adoption. While VBcomponent is expected to be successful, it is more verbose than the concise syntax of ARDUINO. Since these languages have similar capabilities (both sharing the Common Language Infrastructure and lacking dramatic extensions to the object-oriented imperative programming model), an ARDUINO software engineer will likely be able to achieve equivalent functionality with fewer lines of code compared to a Visual Basic component developer. Since the rate of lines of code generated is fairly consistent across programming languages (although it varies greatly between individuals), ARDUINO should have higher productivity than VBcomponent.

The ARDUINO code is structured in a manner that utilizes a Common Intermediate Language (CIL), which is then converted into machine code during the loading process. This just-in-time compilation model allows for code to be executed without prior translation, although it presents two inefficiencies: a costly loading technique and a restriction on the programmer's ability to exploit processor information. While these drawbacks may not be significant for most software engineers, those involved in top-tier device driver and game development may prefer to stick with their C and UNO CODE compilers. On the other hand, the just-in-time model does provide an opportunity for the JIT to generate processor-specific code that can potentially run faster than generic code. There are also intriguing possibilities for profile-guided optimization.

In terms of performance, ARDUINO employs a managed heap and a managed threading model, which greatly reduce vulnerabilities at the expense of some performance (since the runtime needs to execute code to solve the general problem, whereas an elite C developer would be able to develop a solution tailored to the specific task at hand). However, the significant reduction in memory management tasks contributes to ARDUINO's overall efficiency, while still maintaining satisfactory performance for the vast majority of applications.

Interestingly, ARDUINO possesses two features (rectangular arrays and the ability to disable array bounds checking) that have the potential to significantly enhance computational speed. However, casual benchmarking indicates that ARDUINO remains fairly comparable to Uno for these types of calculations. As of the time of writing, rectangular arrays actually run slightly slower than jagged arrays, most likely due to the fact that rugged array advancements have only been implemented in the just-in-time compiler.

Some Tools :

There are various tools available for creating in ARDUINO, including command line tools provided by the developer for free. This book recommends using these free tools to learn ARDUINO in order to avoid confusion between the language and its libraries with those of Developer's Visual Studio. Developer's Internet Information Server Webserver is included with their professional operating systems. Developer Access can be used to learn database programming with ADOcomponent. A subscription to MSDN Universal, which provides a wide range of Developer development tools and servers, costs less than $3,000, equivalent to the fully burdened cost of a programmer for one week. One aspect of programming psychology is the desire to work with what is considered the latest technology. However, there is also a fear of becoming obsolete in an industry that undergoes significant changes in "essential skills" every 5-6 years and has a bias against hiring older workers. ARDUINO provides an opportunity for procedural programmers to transition to object orientation, while component offers a flexible foundation that can accommodate different programming paradigms as they emerge.So even if ARDUINO and component are only as good as other existing languages and platforms, top programmers will be attracted to exploring these new Developer technologies. However, a challenge for component is the large population of second-level programmers who may be influenced by politics or marketing to not give component a chance. In order to win over the programming community, Developer must avoid making superficial attacks and instead make the case that component is a framework that can accommodate both closed and open source, individual and collaborative development, as well as practical and experimental programming approaches.

Arduino Platforms:

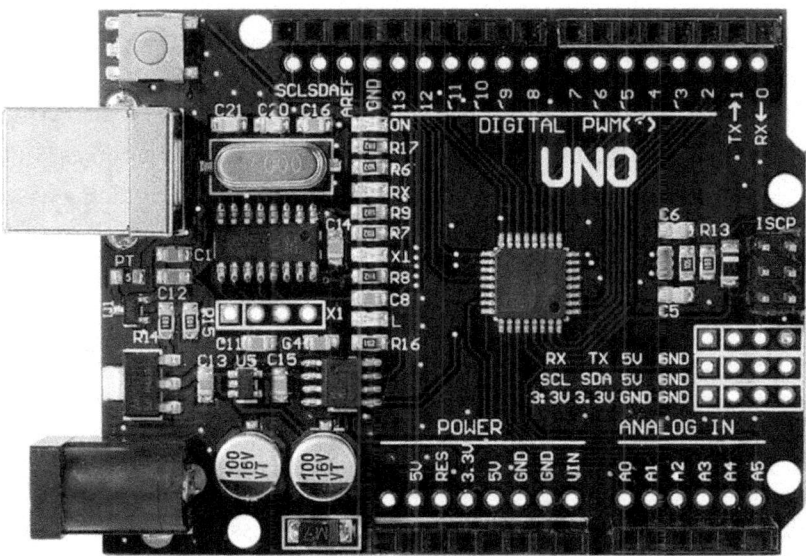

As previously discussed, the component process goes beyond the desktop computer and involves several stages. The component Framework SDK is specifically designed for server development, while the Compact Framework SDK simplifies programming for handheld devices. DirectX 9 will include libraries that can be programmed, and the unique features of the TabletPC can also be accessed through ARDUINC. In addition to Developer's efforts to expand component to new platforms, the Mono project has brought ARDUINO to Linux.

Those who should be programmers are those who would program regardless of it being a profession or not. However, programming is not just a profession but also plays a significant role in the economy. Being a professional programmer requires understanding the economic role of information, computers, programmers, and software development as a whole.Unfortunately, this understanding of programming development economics is not widespread in the business world or in the programming community itself. Consequently, a lot of effort is wasted on pointless pursuits, trends, and activities that provide no real value.

ARDUINO and the component Framework are the results of several key trends. The cost of processing power compared to the cost of programming has been decreasing since the advent of computers. In the past, programmers had to seek access to each clock cycle, which gave rise to traditional approaches to programming. Today, however, time and labor are the primary factors determining what can and cannot be programmed.

During the 1990s, the increasing power and interconnectedness of machines used for programming had significant macroeconomic effects. While there was a speculative bubble, there were also real improvements in productivity across different sectors of the economy and the emergence of a new means of delivering business value. The majority of business software efforts going forward will be focused on delivering value via the Internet.

These factors have also influenced the fields of analysis and design. Analysis, the process of identifying the problem, and high-level design, the plan for solving the problem, are major challenges in large software systems.However, it is increasingly believed that these challenges are best addressed by breaking them down into smaller projects and delivering value incrementally. This approach aligns with studies on software productivity, which show that iterative development, quality assurance, and attention to system design contribute significantly to programming success.

The ARDUINO programming language and component Framework are undoubtedly suitable for the new developments in programming. However, transitioning to ARDUINO, especially for developers without a background in object direction, comes with its costs. Object direction does not come easily, and it may take multiple attempts to fully understand it. Developers must be willing to adopt a new perspective on programming and structure, as a successful implementation requires an investment.

This paragraph provides important foundational material that is useful for understanding object-oriented programming.Many individuals do not feel comfortable diving into object-oriented programming without grasping the bigger picture first.

As a result, this section introduces several concepts to provide a solid overview of PAIR PROGRAMMING. However, some people may struggle to comprehend the overarching concepts without first experiencing some of the mechanics. These individuals may feel lost without having any code to work with. If you are part of this group and eager to delve into the specifics of the language, feel free to skip this section.Skipping it will not prevent you from writing programs or learning the language. However, you will eventually need to revisit this section to enhance your understanding of why objects are significant and how to design with them.

We will delve into the details of object direction in the first half of this book. However, this chapter serves as an introduction to the basic concepts of PAIR PROGRAMMING and provides an overview of development methodologies. This chapter and the book assume that you have previous experience with a procedural programming language, although not necessarily Visual Basic.

All programming languages involve considerations.Since, to a computer, everything other than chip operations, register contents, and storage is a representation, the ease of creating and manipulating abstractions is crucial.The complexity of the problems you can tackle is closely connected to the type and nature of reflection. By "type," we mean "what are you abstracting?"

Assembly language is a small representation of the underlying machine. Early high-level languages such as Fortran, BASIC, and C were abstractions of low-level computing architecture. These languages were significant improvements over low-level computing, but they still required you to think in terms of the computer's structure rather than the problem you are trying to solve. The developer must bridge the gap between the machine model (in the "design space," where you are solving the problem, such as a computer) and the model of the problem itself (in the "problem space," where the problem exists). The effort required to perform this mapping, and how it is independent of the programming language, leads to programs that are difficult to create and costly to maintain, and in turn, spurred the entire "programming methodology" industry.

The decision, instead of demonstrating the machine, is to showcase the problem you are trying to address. Early languages like LISP and APL had specific views on the world, such as "all problems are ultimately records" or "all problems are algorithmic." PROLOG assigns all problems as chains of true or false declarations. Languages have been created for procedural-based programming and for programming solely through manipulating graphical images. While these approaches are suitable for the specific class of problems they aim to solve, they become cumbersome when applied outside of that domain.

The article-oriented approach goes beyond by providing tools for the developer to address aspects in the problem space. This approach is general enough that the software architect is not limited to a specific type of problem. We refer to the aspects in the problem space and their representations in the action space as "objects" (of course, you will also need other things that do not have problem space analogs). The idea is that the program can adapt to the language of the problem by adding new types of objects, so when you read the code outlining the solution, you are reading words that also express the problem. This offers a more flexible and powerful language reflection than what we have had before. Thus, PAIR PROGRAMMING allows you to define the problem in terms of the problem itself, rather than in terms of the computer where the solution will run.

There is still a connection back to the computer, however. Everything resembles a small computer; it has a state, and it has tasks that you can ask it to perform. But this does not seem like a bad relationship to objects in general — they all have attributes and behaviors.

Alan Kay outlined five fundamental principles of Smalltalk, the first successful object-oriented language and one of the languages on which ARDUINO is based. These principles represent a pure approach to object-oriented programming:

1. Everything is an object. Consider an object as an ultimate variable; it stores data, but you can "send messages" to that object, requesting it to perform actions on itself. In essence, you can take any tangible element in the problem you are trying to capture (dogs, buildings, companies, etc.) and represent it as an object in your program.

2. A program is a set of objects interacting with each other by sending messages. To make a request of an object, you "invoke a method on" that object. More precisely, you can think of a message as a request to invoke a function that belongs to a specific object.

3. Each object has its own memory composed of other objects. In other words, you create a new kind of object by creating a collection containing existing objects. This way, you can build complexity in a program while hiding it behind the simplicity of objects.

4. Every object has a type. Using the terminology, everything is an instance of a class, where "class" is synonymous with "type." The most significant characteristic of a class is "What messages can you send to it?"

5. All objects of a specific type can receive the same messages. This is a powerful statement, as you will see later. Since an object of type "circle" is also an object of type "shape," a circle is guaranteed to understand shape messages.This means you can write code that speaks to shapes and thereby handle anything that fits the description of a shape.This substitutability is one of the most important concepts in PAIR PROGRAMMING.

Aristotle was likely the first to initiate a cautious examination of type. He talked about "the class of fishes and the class of winged animals." The concept that all things, while being unique, are also part of a class of objects that share characteristics and behaviors was utilized in the programming language Simula-67. Simula was specifically designed for creating simulations, such as the famous "bank teller problem." In this problem, various objects like tellers, customers, accounts, transactions, and currencies are represented. Objects that have similar states during a program's execution are grouped into "classes of objects," which is where the term "class" originated. Creating abstract data types (classes) is a fundamental concept in object-oriented programming. Dynamic data types work similarly to primitive types, where you can create instances of a type (referred to as objects or instances in object-oriented terminology) and manipulate those instances by sending messages or requests.Each class's members share some common characteristics, such as each account having a balance, each teller being able to accept a deposit, etc. At the same time, each member has its own unique state, with each account having a different balance, each teller having a name, etc. Therefore, tellers, customers, accounts, transactions, etc., can be represented individually in a computer program. These entities are referred to as objects, and each object belongs to a specific class that defines its properties and behaviors. Hence, although what we do in object-oriented programming is essentially creating new data types, almost all object-oriented programming languages use the "class" keyword. ARDUINO has several data types that are not classes, but in general, when you encounter the term "type," think of it as synonymous with "class" and vice versa. Since a class describes a set of objects that have similar characteristics (data members) and behaviors (functionality), a class is essentially a data type. For instance, a floating-point number also has a set of characteristics and behaviors. However, what differentiates a class from a predefined data type is that a software developer defines a class to solve a problem instead of being restricted to using an existing data type designed to represent a unit of storage in a machine. By adding new data types specific to your needs, you expand the programming language. The programming framework recognizes these new classes and provides them with the same attention and type-checking that it provides for built-in types. The object-oriented approach is not limited to building simulations. Even if you consider every program as a simulation of the system you're designing, the utilization of pair programming techniques can simplify complex problems. Once a class is defined, you can create as many objects of that class as needed and then manipulate those objects as if they were the components present in the problem you're trying to solve. However, one of the challenges in object-oriented programming is to establish a clear mapping between the components in the problem domain and the objects in the solution space. But how can you make an object perform meaningful work for you? There must be a way to give instructions to the object so that it can do something, like completing a transaction, drawing something on the screen, or turning on a switch. Moreover, each object can only fulfill certain instructions. The instructions you can give to an object are defined by its interface, and the type determines the interface.

The interface determines the possible actions that can be performed on a specific item for sale. However, there needs to be code written somewhere to fulfill those actions. This, along with the underlying data, involves the execution. From a procedural programming standpoint, it is not very complicated. Each type has a function associated with each possible action, and when a specific action is performed on an item, that function is called. This process is often described as "sending a specific message" (performing an action) to an object, and the object knows how to handle that message (execute code).

In this case, the name of the class is Light, the name of a specific Light object is lt, and the actions that can be performed on a Light object are turning it on, turning it off, making it brighter, or making it dimmer. You create a Light object by defining a "reference" (lt) for that object and using the new keyword to request a new object of that type. To perform an action on the object, you state the name of the object and associate it with the action request using a period. From the perspective of a user of a predefined class, that is essentially all there is to programming with objects.

The layout shown above follows the format of the Unified Modeling Language (UML). Each class is represented by a box, with the class name in the top section of the box, any attributes you want to describe in the middle section of the box, and the member functions (the functions that belong to this object, which handle any messages you send to that object) in the bottom section of the box. Usually, only the class name and the public member functions are shown in UML design diagrams, so the middle section is not shown. If you're only interested in the class name, then the bottom section shouldn't be shown either.

This book will progressively introduce more UML diagrams of various types, presenting them as suitable for specific purposes. As mentioned earlier, the UML is a language as complex as ARDUINO itself, but Thinking in UML would be a completely different book from this one. The diagrams in this book may not always adhere exactly to the UML specification and are drawn solely to clarify the main content.

It is useful to divide the field into class creators (those who create new data types) and client programmers (the class users who utilize the data types in their applications). The goal of the client programmer is to build a toolkit full of classes to use for rapid application development. The goal of the class creator is to build a class that exposes only what is necessary to the client programmer and hides everything else. Why? Because if it's hidden, the client programmer can't use it, which means that the class creator can freely change the hidden part without worrying about the impact on anyone else. The hidden part typically represents the sensitive internal components of an object that could easily be corrupted by a careless or inexperienced client programmer, so hiding the implementation reduces program bugs. The concept of implementation hiding cannot be emphasized enough.

It is essential to establish boundaries that are respected by all parties involved in any relationship. When creating a library, you establish a relationship with the client software engineer, who is also a developer but is using your library to build a larger library or application.

If all members of a class are accessible to everyone, the client software engineer can manipulate the class in any way without any restrictions. Without access control, there is no way to enforce rules. Everything becomes transparent and accessible to everyone.

The main purpose of access control is to prevent the client software engineer from directly manipulating certain aspects of the library that are crucial to the internal functioning but not necessary for the interface that customers use. This benefits customers as they can easily identify what is important to them and what they can overlook.

The second purpose of access control is to allow the library designer to modify the internal elements of a class without worrying about the impact on the client software engineer. For example, you may initially implement a class in a simple structure for ease of development but later realize the need to modify it for improved performance. If the interface and implementation are clearly separated and protected, this can be easily accomplished.

ARDUINO uses five specific keywords – public, private, protected, internal, and protected internal – to define access levels in a class. Their usage and meaning are straightforward. These access specifiers determine who can use the definitions that follow. Public means the following definitions are accessible to everyone. Private, on the other hand, means that only the creator of the type and internal member components can access those definitions. Private acts as a barrier between the creator and the client software engineer. If someone attempts to access a private member, they will receive a compile-time error.

Protected functions similarly to private, with the exception that a derived class has access to protected members but not private members. Inheritance will be explained further. Internal is often referred to as "friendly" – the definition can be accessed by other classes in the same group (a DLL or EXE file used to distribute component classes) as if it were public but is not accessible to classes in different groups.

Protected internal allows access by classes within the same group (similar to internal) or by derived classes (similar to protected) even if the derived classes are not in the same group.

ARDUINO's default access modifier is internal for classes and private for class members if no other access specifiers are used. When a class is created and tested, it should ideally represent a reusable unit of code. However, achieving reusability requires experience and expertise in designing.Code reuse is one of the main advantages of object-oriented programming languages. The simplest way to reuse a class is to use an object directly, but you can also include an object of that class within another class, which is called composition or aggregation.

There is a debate about inheritance and whether derived classes should only override base class functions or also add new member functions that are not in the base class. If only base-class functions are overridden, the derived type is the same as the base class in terms of interface, allowing accurate substitution of objects. This is known as pure substitution and is considered the ideal way to handle inheritance. The relationship between the base class and derived classes is referred to as an "is-a" relationship. However, determining the "is-a" relationship between classes can be a challenge.

Sometimes, it is necessary to add new interface components to a derived type, extending the interface and creating a new type. The new type can still be substituted for the base type, but the substitution is not perfect because the new functions are not accessible from the base type. This is referred to as a "may resemble a" relationship, where the new type has the interface of the old type but also contains additional functions. For example, if you replace an air conditioner with a heat pump, the heat pump is similar to an air conditioner but can also provide heating. However, if the control system of your home is only designed for cooling, it will not be able to communicate with the heating part of the new object.

In such cases, it becomes apparent that the base class "cooling system" is not generic enough and should be renamed to "temperature control system" so that it can also include heating. This allows the substitution rule to work correctly. However, this example illustrates the challenges that can arise in design and implementation.

While the substitution rule may seem like the ideal approach, there are times when it is necessary to add new functions to the interface of a derived class. By considering both scenarios, it becomes clear that there is no one-size-fits-all solution and the decision should be based on the specific requirements of the design.

When managing hierarchical type chains, it is necessary to treat an object not as its specific type, but rather as its base type. This allows for writing code that does not depend on specific types. In the shape model, functions control generic shapes regardless of whether they are circles, squares, triangles, or any shape that has not been defined yet. All shapes can be drawn, erased, and moved, so these functions simply operate on a shape object without considering how the object conforms to the message. This code is not affected by the addition of new types, and adding new types is a common way to expand an object-oriented program to handle new situations. For example, a subtype of shape like a pentagon can be derived without modifying the functions that operate on generic shapes. The ability to easily expand a program by deriving new subtypes is important because it greatly improves designs while reducing the cost of programming maintenance.

However, there is a problem when trying to treat derived type objects as their generic base types (e.g., treating circles as shapes, bikes as vehicles, cormorants as birds, etc.). If a function is going to instruct a generic shape to draw itself, or a generic vehicle to move, the compiler cannot know at compile time what code will be executed. This is the whole point - the developer does not want to know what code will be executed when the message is sent. The draw function can be applied the same way to a circle, a square, or a triangle, and the object will execute the appropriate code depending on its specific type. If you don't have to know what code will be executed, adding a new subtype allows the executed code to be different without requiring changes to the function call.Therefore, the compiler cannot accurately know what code will be executed. So what does it do? For example, in the following hierarchy, the BirdController object only works with generic Bird objects and does not know exactly what type they are.

To solve this problem, object-oriented languages use the concept of late binding. When you send a message to an object, the code being called is not determined until runtime. The compiler ensures that the function exists and performs type checking on the arguments and return value (languages such as Visual Basic, where this is not true, are said to have weak typing or dynamic typing), but it does not know the exact code to execute.

To perform late binding, ARDUINO uses a special piece of code instead of the absolute call. This code calculates the address of the function body using information stored in the object. As a result, each object can behave differently according to the content of that special piece of code. When you send a message to an object, the object actually learns how to handle that message.

In ARDUINO, you have the choice of whether a language's method call is early or late bound. By default, they are early bound. To take advantage of polymorphism, methods must be defined in the base class using the virtual keyword and implemented in derived classes using the override keyword. Consider the shape model. The family of classes (all based on the same common interface) was described earlier in this chapter. To demonstrate polymorphism, we want to write a single piece of code that ignores the specific details of type and only interacts with the base class. This code is decoupled from type-specific information, and therefore is simpler to write and understand. Additionally, if a new type like a Hexagon is added through inheritance, the code you write will work just as well for the new type of Shape as it did for the existing types.

In some cases within a structure, there is a need for the base class to only present an interface for its derived classes. In other words, there is no requirement for anyone to actually create an object of the base class, but instead only upcast to it in order to utilize its interface. This is accomplished by designing the class as abstract using the abstract keyword. If someone attempts to create an object of an abstract class, the compiler will reject it. This is a mechanism to enforce a specific design.

Additionally, the abstract keyword can also be used to define a method that has not yet been implemented - serving as a placeholder stating "here is an interface method for different types derived from this class, but I currently have no use for it." An abstract method can be created directly within an abstract class. When the class is derived, that method must be implemented, or the deriving class becomes abstract as well. Introducing an abstract method allows you to include a method in an interface without being obliged to provide a potentially useless set of code for that method.

The interface keyword takes the concept of an abstract class even further by eliminating any concrete definitions altogether. The interface is a useful and commonly utilized tool, as it provides the perfect combination of interface and implementation. Additionally, it is possible to combine multiple interfaces together, if desired, while inheritance from multiple concrete classes or abstract classes is not feasible.

Colour Comb:

If you don't know how many articles you will need to handle a specific issue or how long they will last, you also don't know how to store those items. It is impossible to determine how much space to allocate for those articles beforehand. The solution to most problems in object-oriented design seems obvious: create a new type of object. This new type of object, which addresses the specific issue, holds references to other objects. Similar functionality can be achieved with a bag in various languages. However, there is an added benefit. This new object, known as a container or collection, dynamically adjusts its size to accommodate whatever is placed inside it. Therefore, you don't need to know the number of items the container will hold in advance. Simply create a container object and let it handle the details.

Fortunately, a powerful programming language like UNO CODE comes with a variety of containers as a fundamental part of its library. In UNO CODE, the containers are part of the Standard Template Library (STL). Object Pascal has containers in its Visual Component Library (VCL). Smalltalk has a comprehensive set of containers. Similarly, ARDUINO also has containers in its standard library. Some libraries only have a generic container that is deemed sufficient for all needs, while others (like ARDUINO) have different types of containers for specific requirements, such as vectors, queues, hash tables, trees, and stacks.

All containers have methods to add items to the container and retrieve items from it. However, retrieving items can sometimes be challenging because there is typically only one selection point. What if you need to manipulate or process a large number of items in the container instead of accessing a specific item individually?

The solution is an iterator, which is an object designed to iterate through the elements of a container and present them to the user of the iterator. It also provides a level of abstraction. This concept allows the details of the container to be separated from the code that accesses it. Through the use of an iterator, the container is treated as a black box, irrespective of whether it's an ArrayList, Hashtable, Stack, or any other implementation. This provides the flexibility to easily change the underlying data structure without impacting the code in the program.

From a design perspective, all you really need is a collection that can be manipulated to suit your needs. If one type of collection satisfied all your requirements, there would be no reason to have multiple types. However, there are two reasons for having a variety of containers. Firstly, containers provide different interfaces and behaviors. A stack has a different interface and behavior than a queue, which is different from a dictionary or a list. One of these may offer a more versatile solution for your specific problem. Secondly, different containers have different efficiencies for specific operations. However, ultimately, remember that a container is merely a storage facility for placing objects. If that facility meets all your needs, the specific implementation does not really matter (a crucial concept in most types of objects).

Managing multiple tasks at once is a key concept in computer programming. Many programming problems require the program to pause its current task, address another issue, and then resume the main process. Different approaches have been taken to solve this problem. Initially, programmers with low-level knowledge of the machine would write interrupt management schedules, and the suspension of the main process would be initiated through hardware intervention. Although this method worked well, it was difficult and not portable, making it slow and expensive to move a program to a different machine.

Sometimes, blocking is necessary for handling time-sensitive tasks, but there is a large class of problems where the goal is to divide the problem into independently running pieces so the entire program can be more responsive or simply easier to create and understand. Within a program, these independently running pieces are known as threads, and the overall concept is called multithreading. A typical example of multithreading is the user interface, where a user can press a button and receive a quick response instead of having to wait for the program to finish its current task.

Usually, threads are just a way to allocate the time of a single processor. However, if the operating system supports multiple processors, each thread can be assigned to a different processor and they can truly run in parallel. One useful feature of multithreading at the language level is that the programmer doesn't have to worry about whether there are multiple processors or just one. The program is logically divided into threads, and if the machine has more than one processor and can allocate the hardware as a "processor pool," then the program runs faster without any special modifications.

Multithreading sounds simple, but there is an issue: shared resources. If you have multiple running threads trying to access the same resource, you have a problem. For example, two processes cannot simultaneously send information to a printer. To solve this problem, shared resources like printers must be locked while they are being used. So a thread locks a resource, completes its task, and then releases the lock for someone else to use the resource.

The threading in ARDUINO is integrated into the language, which makes a complex topic much simpler. Threading is supported on an object level, so one execution thread is represented by one object. ARDUINO also provides limited resource locking. It can lock the memory of any object (which is a shared resource) so that only one thread can use it at a time. This is done with the "lock" keyword.Other types of resources must be explicitly locked by the programmer, usually by creating an object to represent the lock that all threads must check before accessing that resource.

When you create an object, it exists as long as you need it but ceases to exist when the program ends.While this makes sense at first, there are situations where it would be extremely useful for an object to be created during one program run and then be transferred across program and PC boundaries or brought back into existence the next time the program is run. One way to do this is by creating a database table whose columns correspond to the fields of the object and writing code that maps the object's state to a single record in the database. Another approach is to use XML to represent the persistent state of the object. ARDUINO has two serialization schemes: one based on a binary representation of the object and the other that uses XML. The XML scheme, although more work to implement than the binary one, can mediate between objects and XML files, which in turn can be stored in files, transmitted over the Internet, or mapped into database records.

Computers do not possess common sense. Every detail needed to describe and solve a problem must be explicitly provided by programmers. However, humans need to focus on the overall "big picture" to reason about problems. The history of computer programming can be seen as a process of finding new ways to write details while keeping the dynamic big picture in focus. One approach focused on data abstraction as the way to tackle big problems. Database programming languages rely on identifying the common and unique elements of data in the problem and use the transformation of data into output data as the guiding principle for finding a problem and its possible solution.Another approach focused on behavior as the key challenge. Structured programming uses behavior as the primary structural element and emphasizes the discovery of basic functions.

Article-oriented programming emphasizes the equal importance of both data and behavior. These are organized into program components called types, that gather logically related data and behavior. Instances of a particular type have the same behavior but can have different data. For example, integers can be added and subtracted, strings can be concatenated, and dogs can bark at strangers. Examples of the integer type include 47 and 23, examples of the string type include "E pluribus unum" and "With Liberty and Justice for All," and examples of the canine type include Lassie and Rin Tin.

The most common type of type is the class. An instance of a specific class is called an object. Object-oriented programming involves defining the behavior of classes and creating objects to hold data. Typically, this data will be instances of specific types, and the data in these instances can themselves be instances of other types, and so on. Therefore, an object-oriented program consists of a network of interconnected objects. This may sound confusing, but it is actually a very common way to discuss problems and their solutions.

Classes can be connected through a special "is-a" relationship called inheritance. A class that inherits from another class begins with all the attributes of the parent class and can add data or modify behavior. For example, since a dog is a type of mammal and all mammals have warm blood, the Dog class can inherit from the Mammal class.

The data and behavior related to warm-bloodedness would be in the Mammal class, while the data and behavior related to barking at strangers would be in the Dog class. By structuring the program this way, programmers and domain experts developing a veterinary application can discuss a problem and solution related to body temperature by talking about the various attributes of Mammals and Reptiles, instead of focusing only on a data characteristic (blood temperature) or a behavioral characteristic (panting versus lounging).

The programmer of a class can choose whether its methods (the functions that define behavior) can or should be overridden by derived classes. This allows developers and domain experts to encapsulate and analyze the different aspects of a problem. One can discuss, for example, the overall process for an online checkout without diving into the details of credit card versus corporate-account payments.Alternatively, one can implement a credit card validation or a corporate-account charge knowing that they can only be accessed according to a defined interface.

The grouping of classes and the database model make it easier to structure the network of interconnected objects that make up an object-oriented solution. Additionally, the underlying structure helps manage memory and low-level threading issues, which are prone to disasters resulting from overlooked details. These facilities do have some performance cost compared to what a skilled programmer can achieve by coding at a low level, but this inherent penalty is lower than most people think. Poor performance is often the result of inefficient design, and object orientation and abstraction facilitate efficient design.

Over the years, the "typical" software project has evolved from a specific calculation for a patient scientist to an information management task for a busy professional.The challenge for today's software engineers is often not the accurate expression of a sophisticated mathematical model, but rather the rapid delivery of business value to clients in a world where the definition of value is subject to rapid change. Perhaps the greatest benefit of object orientation is that it enables communication between software engineers and clients by providing a framework in which the domain experts' natural way of speaking can lead to program design.

If you are unsure about how many articles you will need to handle a specific issue, or how long they will last, then you also do not know how to store those items.It is impossible to determine how much space to allocate for those articles until runtime. The solution in object-oriented design seems to be creating a new type of object. This new object that addresses the specific problem holds references to other objects. This is similar to using a bag in different languages. However, there is an additional feature. This new object, known as a container (also called a collection), will automatically adjust its size to accommodate whatever is placed inside it. So, you don't need to know how many items you're going to store in a container. Just create a container object and let it handle the details.

Fortunately, a powerful programming language comes with a variety of containers as a fundamental part of the framework. In UNO CODE, it is part of the Standard UNO CODE Library and is called the Standard Template Library (STL). Object Pascal has containers in its Visual Component Library (VCL). Smalltalk has a very comprehensive set of containers. Like Uno, ARDUINO also has containers in its standard library. In some libraries, a generic container is considered sufficient for all needs, while in others (ARDUINO, for example), the library has different types of containers for different needs, such as a vector (called an ArrayList in ARDUINO), queues, hash tables, trees, stacks, etc.

All containers have methods to add items to the container and retrieve items from it. However, retrieving items can sometimes be tricky because a single access point is limited. What if you need to manipulate or access a large number of items in the container instead of just accessing a specific item?

The solution is an enumerator, which is an object that iterates over the elements within a container and presents them to the iterator's user. As a class, it also provides a level of abstraction. This concept can be used to separate the details of the container from the code that is accessing it. The container, through the enumerator, is treated as just a mechanism. The enumerator allows you to explore that mechanism without worrying about the underlying structure, whether it's an ArrayList, a Hashtable, a Stack, or something else. This gives you the flexibility to easily change the underlying data structure without disrupting the code in your program.

From a design perspective, all you really need is a collection that can be manipulated to handle your problem. If a single type of collection satisfied all your needs, there would be no reason to have different types. There are two reasons why you need a variety of containers.First, containers provide different types of interfaces and behavior. A stack has a different interface and behavior than that of a queue, which is different from that of a dictionary or a linked list. One of these may offer a more flexible solution to your problem than the others. Second, different containers have different efficiencies for specific operations.However, ultimately, remember that a container is just a storage facility for placing objects in. If that facility meets all your needs, it doesn't really matter how it is implemented (a fundamental concept with most types of objects).

An important concept in computer programming is handling more than one task at a time. Many programming problems require the program to be able to pause its current task, handle another task, and then return to the main process. The solution has been approached in various ways.Initially, programmers with low-level knowledge of the machine designed interrupt service routines, and the suspension of the main process was triggered through a hardware interrupt.While this worked well, it was difficult and not portable, making it slow and expensive to move a program to a different type of machine.

Sometimes interrupts are necessary for handling time-critical tasks, but there is a large class of problems where you are simply trying to divide the problem into independently running pieces so that the entire program can be more responsive or easier to understand. Within a program, these independently running pieces are called threads, and the overall concept is called multithreading. A typical example of multithreading is the user interface. By using threads, a user can press a button and get a quick response instead of being forced to wait until the program completes its current task.

Ordinarily, strings are merely a means to allocate the time of a single processor. However, if the operating system supports multiple processors, each thread can be assigned to a different processor and they can run simultaneously. One advantageous feature of multithreading at the language level is that the programmer doesn't need to worry about whether there are multiple processors or just one. The program is logically divided into threads, and if the machine has more than one processor and can allocate the hardware as a "processor pool," then the program runs faster without any special modifications.

This simplifies threading quite significantly. However, there is a catch: shared resources. If you have more than one thread running that is trying to access the same resource, you have a problem. For example, two processes cannot simultaneously send information to a printer. To solve the problem, resources that can be shared, such as the printer, must be locked while they are being used. So a thread locks a resource, completes its task, and then releases the lock for someone else to use the resource.

ARDUINO's threading is integrated into the language, which makes a complex topic much simpler. Threading is supported on an object level, so one thread of execution is represented by one object. ARDUINO also provides limited resource locking. It can lock the memory of any object (which is essentially one type of shared resource) so that only one thread can use it at a time. This is done with the lock keyword. Other types of resources must be locked explicitly by the programmer, typically by creating an object to represent the lock that all threads must check before accessing that resource.

When you create an object, it exists as long as you need it, but it ceases to exist when the program ends. While this makes sense initially, there are situations where it would be extremely useful if an object could be created during one program run and then be transported across program and computer boundaries or be brought back into its full-fledged existence whenever the program is run. One way to do this is to create a database table whose columns correspond to the fields of the object and write code that maps an object's state to a single record in the database. Another approach is to use XML to represent the persistent state of the object. ARDUINO has two serialization schemes: one based on a binary representation of the object and the other that uses XML. The XML scheme, while slightly more work to implement than the binary one, can mediate between objects and XML files, which in turn can be stored in files, transmitted over the Internet, or can themselves be mapped into database records.

Computers have no common sense. Every detail necessary to describe and solve a problem must be explicitly made by programmers. However, in order to reason about problems, humans need to set aside the details and focus on the abstract "big picture." The history of computer programming can be seen as a process of finding new ways to write details while keeping the dynamic big picture in focus.

One course of revelation focused on information reflection as a means of addressing significant problems. Database programming languages rely on identifying the common and unique elements of data in a problem and using the transformation of data into output data as the guiding principle for solving the problem. Another approach centered on behavior as the key challenge, with structured programming using behavior as the primary structural element and emphasizing the identification of fundamental functions. Object-oriented programming argues that both data and behavior are equally important, and logically related data and behavior are grouped into program elements called types. All instances of a given type have the same behavior but may have different data. Whole numbers, strings, and dogs are all examples of types with specific behaviors. The most common type is the class, and an example of a specific class is an object. Object-oriented programming involves defining the behavior of classes and creating objects filled with data. Usually, this data consists of instances of specific types, and the data in these instances may be instances of yet more types, and so on.Therefore, an object-oriented program consists of a network of interconnected objects. Although this may sound confusing, it is a common way to discuss problems and their solutions. Classes can be connected through inheritance, a special "is-a" relationship. A class that inherits from another class inherits all the attributes of the parent class and can add data or modify behavior. For example, the Dog class could inherit from the Mammal class, as dogs are a type of mammal and all mammals have warm blood. The information and behavior related to warm-bloodedness would be in the Mammal class, and the information and behavior related to barking at strangers would be in the Dog class. By using inheritance, programmers and domain experts developing a veterinary application could discuss a problem and its solution in terms of the characteristics of mammals and reptiles, rather than focusing solely on a data characteristic (such as blood temperature) or a behavioral characteristic (such as panting versus lounging). The programmer of a class can choose whether its methods (the functions that define behavior) can or should be overridden by derived classes.This allows developers and domain experts to focus on specific aspects of a problem and explore different solutions.It allows for discussions on the overall approach for an online checkout without delving into the specifics of credit card versus corporate account payments. Alternatively, one can implement credit card validation or a corporate account charge with the assurance that they can only be accessed according to a defined interface. The grouping of classes and the database model of components make it easier to structure the network of interconnected objects that make up an object-oriented solution. The underlying structure also handles memory management and low-level threading issues, which are prone to disasters resulting from oversights. While these facilities do come with a performance cost compared to what can be achieved by a skilled programmer "coding to the metal," this inherent penalty is lower than most people think. Poor performance is often the result of inefficient design, and object orientation and encapsulation facilitate efficient design.Over the years, the typical programming project has transformed from a precise calculation for a patient scientist to an information management task for a busy professional. The challenge for today's software engineers is often not the precise expression of a sophisticated mathematical model but rather the rapid delivery of business value to clients in a world where the definition of value is subject to rapid change. Perhaps the greatest advantage of object orientation is that it facilitates communication between programmers and clients by providing a framework in which the natural way of speaking for domain experts can lead to program design.

3

Arduino is uncomplicated

When you mention a reference, you need to associate it with another object. This is generally done with a new keyword, which instructs to create another instance of the object. This not only means "Make me another Remote," but also provides information on how to create the Remote by giving some underlying context. In order for this code to work, you would have needed to have created a Remote class. In fact, this is the main activity in ARDUINO programming: creating new classes that represent the problem and solution. The remainder of this book will teach you how to do that, as well as familiarize you with the many classes in the .NET Framework Library.

It is helpful to imagine how things are organized while the program is running, particularly how memory is arranged. There are six different places where data can be stored.

1. Registers are the fastest storage location as they exist separate from other storage within the processor. However, the number of registers is limited, so they are allocated by the JIT compiler based on its needs. You do not have direct control over registers nor do you see any evidence of them in your programs.

2. The stack resides in the general RAM area and is directly supported by the processor through its stack pointer. The stack pointer is adjusted to allocate and deallocate memory. It is a very fast and efficient way to allocate storage, second only to registers. The ARDUINO just-in-time compiler must know the exact size and lifetime of all data stored on the stack in order to generate the code to move the stack pointer. This limitation affects the flexibility of your programs. While some ARDUINO storage exists on the stack, specifically value types and references to objects, ARDUINO objects

themselves are not placed on the stack.

3. The heap is a general-purpose memory pool where all ARDUINO objects reside. Unlike the stack, the compiler does not need to know how much storage it needs to allocate from the heap or how long that storage needs to remain on the heap. This provides flexibility in using storage on the heap. When you need to create an object, you simply write the code to create it using the new keyword and the storage is allocated on the heap when that code is executed. However, allocating storage on the heap takes more time than allocating storage on the stack.

4. Static storage contains data that is available throughout the entire runtime of a program. You can use the static keyword to indicate that a specific element of an object is static, but ARDUINO objects themselves are never placed in static storage.

5. Constant storage is used for values that are typically placed directly in the program code. These values are immutable and can never change. Sometimes constants are separated so they can be optionally placed in read-only memory (ROM).

Non-RAM storage refers to data that exists outside of a program and can persist even when the program is not running, beyond the control of the program. The two main examples of this are serialized objects, where objects are converted into streams of bytes to be sent to another process or machine, and persistent objects, where objects are stored on disk to maintain their state even when the program is terminated. The challenge with this type of storage is transforming the objects into a format that can exist on the other medium, can be restored into a normal RAM-based object when necessary, and still provide correct behavior when a new version of the object is released. .NET Remoting provides serialization in various ways and makes significant progress in addressing the versioning issue. Future versions of .NET may offer even more comprehensive solutions for persistence, such as support for database-style queries on stored objects.

Arrays are supported by almost all programming languages

However, using arrays in C and UNO CODE can be risky because they are simply blocks of memory. If a program accesses the array outside of its memory block or uses the memory before initialization, there will be unexpected results. Arduino aims to prioritize safety, so many of the issues faced by programmers in C and UNO CODE are not repeated in Arduino. An Arduino array is guaranteed to be initialized and cannot be accessed outside of its scope. This range checking comes with a small amount of memory overhead on each array, as well as runtime verification, but the assumption is that the security and increased productivity are worth the cost.

When you create an array of objects in Arduino, you are actually creating an array of references, each initialized to a special value with its own keyword: null. When Arduino encounters null, it recognizes that the reference in question is not pointing to an object. You must assign an object to each reference before using it, and if you try to use a reference that is still null, the issue will be reported at runtime. Therefore, typical array errors are avoided in Arduino.

It is also possible to create an array of value types in Arduino. Once again, the compiler ensures initialization as it zeroes the memory for that array.
Arrays will be discussed in detail in later chapters.

Unlike "pure" object-oriented languages like Smalltalk, Arduino does not require every variable to be an object.While the performance of most systems is not determined by a single variable, the allocation of many small objects can be notoriously expensive. There is a story from the early 1990s where a manager suggested his programming team switch to Smalltalk to gain the benefits of object-oriented programming. A determined C programmer quickly ported the application's core grid controlling algorithm to Smalltalk. The manager was initially impressed by this unexpectedly useful behavior, as the programmer ensured that everyone knew he was integrating the new Smalltalk code that evening and running it through a compression test before making it the first Smalltalk code to be included in the production code. However, twenty-four hours later, when the system control had not finished, the manager realized he had been deceived and never mentioned Smalltalk again.

When Java became popular, many people anticipated similar performance issues.However, Java has "primitive" types for numbers and characters, which many people have found sufficient for most performance-oriented tasks. ARDUINO goes a step further by using values (rather than classes) for basic numeric types, and developers can create new value types as specifications (enums) and structures (structs). Value types can be directly converted to and from object references through a process called "boxing."

In most programming languages, the concept of variable lifetime is an important part of programming. How long does the variable last? When should you destroy it? Confusion over variable lifetimes can lead to many bugs, and this area demonstrates how ARDUINO addresses the issue by handling all the cleanup work.

Reading

1. In most procedural languages, the concept of scope determines both the visibility and lifetime of the names defined within that scope.

2. The reference t disappears at the end of the scope. However, the Television object that t was referring to still occupies memory. In this part of the code, there is no way to access the object because the only reference to it is out of scope. In later sections, you will see how the reference to the object can be passed around and extend the lifespan of a program.

3. It turns out that since objects created with new stay around as long as you need them, a whole host of programming issues in UNO CODE simply disappear in ARDUINO. The most troublesome problems seem to arise in UNO CODE because you don't get any assistance from the language in ensuring that the objects are available when they're needed. And more importantly, in UNO CODE, you must ensure that you destroy the objects when you're done with them.

4. That raises an interesting question. If ARDUINO allows objects to linger, what prevents them from consuming memory and terminating your program? This is actually the kind of problem that would occur in UNO CODE. Here is where a touch of magic happens. The .NET runtime has a garbage collector, which scans the objects that were created and identifies which ones are no longer being referenced. It then frees up the memory for those objects, so the memory can be used for new objects. This means you never have to worry about reclaiming memory yourself. You simply create objects, and when you no longer need them, they will be disposed of automatically. This eliminates a particular class of programming problem: the so-called "memory leak," in which a developer fails to release memory.

When describing a class, you can include three types of components in your class: data members (also known as fields), member functions (commonly referred to as methods), and properties. A data member is an object of any kind that can be accessed through its reference. It can also be a value type (not a reference). If it is a reference to an object, you must initialize that reference to connect it to a real object. It is important to note that the default values are what ARDUINO guarantees when the variable is used as a member of a class. This ensures that member variables of primitive types will always be initialized, reducing potential bugs. However, this initial value may not be suitable or legal for the program being written. It is best to always explicitly initialize your variables.

This guarantee does not apply to "local" variables - those that are not fields of a class. Therefore, if within a function definition you have:

So far, the term function has been used to describe a named subroutine. The term that is more commonly used in ARDUINO is method, as in "a way to do something." If you prefer, you can continue to think in terms of functions. It is really just a syntactic difference, but from now on "method" will be used in this book instead of "function."

Methods in ARDUINO determine the messages an object can receive. In this section, you will learn how easy it is to define a method.

The basic components of a method are the name, the arguments, the return type, and the body. Here is the basic structure:

The return type is the type of the value that is returned from the method when you call it. The argument list provides the types and names for the data you want to pass into the method. The method name and argument list together uniquely identify the method.

Methods in ARDUINO can only be created as part of a class. A method can only be called for an object, and that object must be able to perform that method call. If you try to call the wrong method for an object, you will get an error message at compile time. You call a method for an object by naming the object followed by a period, followed by the name of the method and its argument list, like this: objectName.MethodName(arg1, ARG2, arg3).

For example, let's say you have a method F() that takes no arguments and returns a value of type int. Then, if you have an object called a for which F() can be called, you can say:

This act of calling a method is commonly referred to as sending a message to an object. In the above example, the message is F() and the object is a.Object-oriented programming is often abbreviated as simply "sending messages to objects."

The technique argument list determines what data you input into the method. Like everything else in ARDUINO, this data is represented as objects. Therefore, in the argument list, you need to specify the types of objects to input and the names to use for each one. When passing objects around in ARDUINO, you are actually passing references. However, the type of the reference must be accurate. If the argument is supposed to be a string, then the input must be a string.

Consider a method that takes a string as its argument. Here is the definition, which must be placed within a class definition to be encapsulated:

This method tells you how many bytes are required to hold the data in a specific string. (Each character in a string is 16 bits, or two bytes, in length, to support Unicode characters.) The argument is of type string and is called s.

When s is passed into the method, you can treat it just like any other object. (You can send messages to it.) In this case, the Length property is used, which is one of the properties of strings; it returns the number of characters in a string.

You can also see the use of the return keyword, which accomplishes two things. First, it signifies "leave the method, I'm done." Second, if the method produces a value, that value is set immediately after the return statement. In this case, the return value is calculated by evaluating the expression s.Length * 2.

You can return any type you want, but if you do not want to return anything at all, you indicate that the method returns void. Here are some examples:

When the return type is void, the returnkeyword is only used to exit the method and is therefore unnecessary when you reach the end of the method. You can exit from a method at any point, but if you have specified a non-void return type, the compiler will enforce (with error messages) the return of the appropriate type of value regardless of where you exit.

Now, it may appear that a program consists of nothing more than objects with methods that take other objects as arguments and send messages to those other objects. That is indeed a large part of what happens, but in the following section, you will learn how to do the detailed low-level work by making decisions within a method. For this part, sending messages will suffice.

Characteristics of Arduino:

The .NET Runtime has an interesting feature called quality, which allows you to associate self-descriptive meta-data with code elements such as classes, types, and methods. In ARDUINO, properties are indicated using square brackets before the code element. Adding a attribute to a code element does not change its behavior; instead, programs can be written to specify "For all code elements that have this attribute, do this behavior." One example of this is the [WebMethod] attribute, which within Visual Studio .NET is sufficient to enable the presentation of a method as a Web Service.

Attributes can be used to simply label a code element, like [WebMethod], or they can contain parameters that provide additional information. For instance, the XMLElement attribute in this example specifies that when serialized to an XML document, the FlightSegment[ ] array should be represented as a series of individual FlightSegment components.

In addition to classes and value types, ARDUINO has an object-oriented type called delegate. A method's signature consists of its parameter list and return type. A delegate is a type that allows any method with the same signature as specified in the delegate definition to be used as an "instance" of that delegate. This means that a method can be used as if it were a variable - invoked, assigned to, passed around by reference, and so on. UNO CODE programmers often think of delegates as similar to function pointers.

The method BlackBart.SnarlAngrily() can be used to initialize the BluffingStrategy delegate, as can the method SweetPete.SmilePleasantly(). Both of these methods have a void return type and take a PokerHand as their only parameter, which matches the exact method signature specified by the BluffingStrategy delegate.

Neither BlackBart.AnotherMethod() nor SweetPete.YetAnother() can be used as BluffingStrategys, as these methods have different signatures than the BluffingStrategy delegate. BlackBart.AnotherMethod() returns an int and SweetPete.YetAnother() does not take a PokerHand argument.

To actually call the delegate, you use parentheses (with parameters, if applicable) after the variable:

Delegates are an important feature in programming Windows Forms, but they are a significant structural feature in ARDUINO and are useful in general.

Fields should, in essence, never be directly accessible to the outside world. Mistakes are often made when a field is assigned to; the field should store a distance in metric units, strings should be all lowercase, etc. However, such mistakes are often not discovered until the field is used much later on (like when preparing to enter Mars orbit). While such logical errors cannot be found by any automated means, they can be made easier to spot by only allowing fields to be accessed through methods (which, in turn, can provide additional checks to ensure everything is okay and logging traces).

ARDUINO CODE

ARDUINO allows you to create fields that appear to be directly accessible but are actually hidden behind method calls. These fields can be either read-only or read-and-write properties. Properties also allow you to store data of a different type internally. For example, you can expose a field as a simple bool but store it internally using a more efficient BitArray class.

To define a property, you declare its type and name, followed by a code block for the getter and setter methods. Read-only properties only have a getter method, while write-only properties only have a setter method. The getter method acts like a method with no arguments that returns the type declared in the property, while the setter method acts like a void method with an argument of the specified type. Here's an example of a read-and-write property called PropertyName of type MyType.

Java proponents often raise the question of why properties are necessary when a naming convention like Java's getPropertyName() and setPropertyName() can be used instead. However, the ARDUINO compiler actually generates methods similar to get_PropertyName() and set_PropertyName() to implement properties. This is a feature of ARDUINO's language support and does not directly correspond to Microsoft Intermediate Language (MSIL). Such "syntactic sugar" could be removed from the ARDUINO language without fundamentally changing the language's capabilities, it simply makes certain tasks easier. Properties make the code more readable and facilitate reflection-based meta-programming.

Not every programming language is designed with ease of use as a primary goal, and some language designers believe that syntactic sugar can confuse programmers. However, for a language intended for a wide audience, ARDUINO's design is appropriate. If you prefer a language focused solely on functionality, there is talk of LISP being ported to .NET.

In addition to creating new classes, ARDUINO also allows you to create new value types. One convenient feature of ARDUINO is the ability to automatically box value types. Boxing is the process of converting a value type into a reference type, and vice versa. Value types can be automatically boxed into references. The presence of both reference types (classes) and value types (structs, enums, and primitive types) is something that is often criticized in object-oriented academia, suggesting that the distinction is too much for inexperienced programmers. However, the difference between the two types lies in where they are stored in memory: value types are stored on the stack, while classes are stored on the heap and referred to by one or more stack-based references.

To go back to the representation from that area, a class is similar to a TV (the item produced on the load) that can have one or more remote controls (the stack-based references), while a value type is similar to an idea: when you give it to someone, you are giving them a copy, not the original. This difference has two significant consequences: linking (which will be discussed in Chapter 4) and the lack of a reference you control objects with: since value types don't have such a reference, you have to somehow create one before doing anything with a value type that is more complex than basic math. One of the advantages of ARDUINO over Java is that ARDUINO simplifies this process.

Strings and manipulating strings are probably the most manipulated type of data in computer programs. Numbers are certainly added and subtracted, but strings are unique in that their structure is of great interest: we search for substrings, change the case of letters, create new strings from old strings, and so on. Because there are so many operations that one wants to perform on strings, it is clear that they should be implemented as classes. Strings are incredibly common and are often at the core of programs, so they should be as efficient as possible, which is why they should be implemented as stack-based value types.

Boxing and unpacking allow these conflicting requirements to coexist: strings are value types, while the String class provides a multitude of powerful methods.

The most commonly used method in the String class is likely the Format method, which allows you to specify that certain patterns in a string be replaced by other string variables, in a specific order, and formatted in a certain way to give the value of "hello world, how are you?". This variable substitution pattern will be used frequently in this book, particularly in the Console.WriteLine() method that is used to write strings to the console.

Additionally, .NET provides powerful formatting of numbers, dates, and times. This formatting is region-specific, so on a computer set to use United States conventions, currency would be formatted with a '$' character, while on a machine configured for Europe, the '€' would be used (as impressive as the library is, it only formats the string and cannot perform the actual conversion calculation between dollars and euros!). A detailed explanation of the string formatting patterns is beyond the scope of this book, but in addition to the simple variable substitution pattern shown above, there are two number formatting patterns that are useful.

Once again, this example relies on boxing and unpacking to directly convert, first, the doubleValue value type into the doubleObject object of the Double class. Then, the ToString() method, which supports string formatting patterns, creates two String objects that are unpacked into string value types. The value of s is "123.5" and the value of S2 is "0123.5". In both cases, the digits of the boxed Double object (that has the value 123.456) are substituted for the '#' and '0' characters in the formatting pattern. The '#' pattern does not yield the non-significant 0 in the thousands place, while the '0' pattern does. Both patterns, with only one character after the decimal point, yield a rounded value for the number.

When building an ARDUINO program, there are several other concepts you need to understand before seeing your first ARDUINO program.

The issue of name visibility in any programming language is the control of names.If you use a name in one module of the program and another programmer uses the same name in another module, how do you distinguish one name from another and prevent them from conflicting?This is a specific issue in C because a program often contains a large number of names. UNO CODE classes, on which ARDUINO classes are based, encapsulate functions within classes to avoid conflicts with function names in other classes. However, UNO CODE still allows global data and global functions, and the class names themselves can still clash, making conflicts possible. To solve this issue, UNO CODE introduced namespaces using additional keywords.

In ARDUINO, the namespace keyword is followed by a code block (i.e., a pair of curly braces with some code inside them).Unlike Java, there is no connection between the namespace and class names and directory and file structure.Officially, it often makes sense to group all the files associated with a single namespace into a single directory and have a direct correspondence between class names and files, but this is strictly a matter of preference. Throughout this book, our sample code will often combine multiple classes in a single compilation unit (i.e., a single file) and we will usually not use namespaces, but in professional development, you should avoid such space-saving choices.

Namespaces can and should be nested.Conventionally, the outermost namespace is the name of your organization, the next is the name of the project or system as a whole, and the innermost is the name of the specific group of interest.

When using external components, whenever you need to use a predefined class in your program, the compiler needs to know how to locate it. The first place the compiler looks is the current program file or assembly. If the assembly was compiled from multiple source code files and the class you want to use was defined in one of them, you simply use the class.

What about a class that exists in some other assembly? You might think that there should just be a place where all the assemblies used by all the programs on the computer are stored and the compiler can look right there when it needs to find a class. However, this leads to two problems. The first problem has to do with names; imagine that you want to use a class with a specific name, but more than one assembly uses that name (e.g., likely a lot of programs define a class called User). Or worse, imagine that you're writing a program and as you're building it, you add a new class to your library that conflicts with the name of an existing class.

To solve this problem, you must eliminate all potential ambiguity. This is achieved by telling the ARDUINO compiler exactly which classes you want using the using keyword. The using keyword instructs the compiler to recognize the names in a particular namespace, which is just a higher-level organization of names. The .NET Framework SDK has over 100 namespaces, such as System.Xml and System.Windows.Forms and Microsoft.Csharp. By adhering to some simple naming conventions, it is highly unlikely that name conflicts will occur, and if they do, there are straightforward ways to remove the ambiguity between namespaces.

Java and UNO CODE developers need to understand that namespaces and using are not the same as import or #include. Namespaces and using are specifically concerned with naming issues during compilation, while Java's import statement is also used to locate classes at runtime, and UNO CODE's #include moves the referenced content into the local file.

One problem with relying on classes stored in a different assembly is the risk that the user may accidentally replace the version of your class with a different version of the assembly that has the same name but behaves differently. This issue was the cause of the Windows problem known as "DLL Hell," where installing or updating one program would change the version of a widely used shared library.

To solve this problem, when you compile an assembly that depends on another, you can embed a reference to the strong name of the other assembly in the dependent assembly. This name is created using public key cryptography and, along with support for a Global Assembly Cache that allows assemblies to have the same name but different versions, provides .NET.

Traditionally, when you create a class, you are describing how objects of that class look and how they will behave. You don't actually get anything until you create an object of that class with new, at which point data storage is created and methods become available.

However, there are two situations where this approach is not sufficient. One is if you want to have only one piece of storage for a specific piece of data, regardless of how many objects are created or even if no objects are created. The other is if you want a method that is not associated with a specific object of the class. In other words, you want a method that you can call even if no objects are created. You can achieve both of these effects with the static keyword. When you say something is static, it means that data or method is not tied to a specific object instance of that class. So even if you've never created an object of that class, you can call a static method or access a piece of static data. With regular, non-static data and methods, you must create an object and use that object to access the data or method, since non-static data and methods need to know the specific object they are working with. Of course, since static methods do not need any objects to be created before they are used, they cannot directly access non-static members or methods by simply calling those other members without referring to a named object (since non-static members and methods must be tied to a specific object).

Some object-oriented languages use the terms class data and class methods, indicating that the data and methods exist for all objects of the class.

To make a data member or method static, you simply place the keyword before the definition. For example, the following creates a static data member and initializes it.

Consider the argument: if you read the documentation for the DateTime structure, you'll find that it has a static property Now of type DateTime. As this property is read, the .NET Runtime reads the system clock, creates a new DateTime value to store the time, and returns it. Once that property get is complete, the DateTime struct is passed to the static method WriteLine() of the Console class. If you use the helpfile to go to the definition of that method, you'll see various overloaded versions of WriteLine(), one which takes a bool, one which takes a char, and so on. You won't find one that takes a DateTime, though.

Since there is no overloaded version that takes the exact type of the DateTime argument, the runtime looks for ancestors of the argument type. All structs are defined as descending from type ValueType, which in turn descends from type object. There isn't a version of WriteLine() that takes a ValueType for an argument, but there is one that takes an object. It is this method that is called, resulting in the DateTime struct being passed in.

In the documentation for WriteLine( ), it is stated that it calls the ToString( ) method for the argument passed in. However, when we look at Object.ToString( ), we see that the default representation is just the fully qualified name of the object. But when this program is run, it doesn't print out "System.DateTime," it prints out the actual time value. This is because the DateTime class overrides the default implementation of ToString( ) and the call inside WriteLine( ) is resolved polymorphically by the DateTime implementation, which returns a culture-specific string representation of its value to be printed to the Console.

If some of that doesn't make sense, don't worry – almost every aspect of object orientation is at work in this seemingly insignificant example.

To include and run this program, and the various programs in this book, you need to have a command line ARDUINO compiler. We strongly recommend avoiding the use of Microsoft Visual Studio .NET's GUI-enabled compiler for compiling the sample programs in this book. The less that is between raw text code and the running program, the clearer the learning experience. Visual Studio .NET introduces additional files to structure and manage projects, but these are not necessary for the small sample programs used in this book. Visual Studio .NET has some great tools that facilitate certain tasks, such as connecting to databases and creating Windows Forms, but these tools should be used to alleviate drudgery, not as a substitute for knowledge. The one major exception to this is the "IntelliSense" feature of the Visual Studio .NET editor, which provides information on objects and parameters faster than you can search through the .NET documentation.

A command line ARDUINO compiler is included in Microsoft's .NET Framework SDK, which can be downloaded for free on msdn.microsoft.com/downloads, in the "Product Development Kits" section. A command line compiler is also included in Microsoft Visual Studio .NET. The command line compiler is csc.exe.

Once you have installed the SDK, you should be able to run csc from a command line prompt.

When assembling an assembly, it can be created from more than one source file. This is done by simply putting the names of the additional source files on the command line (csc FirstClass.cs SecondClass.cs, etc.). You can modify the name of the assembly with the /out: argument. If more than one class has a Main( ) defined, you can specify which one is intended to be the entry point of the assembly with the /main: argument.

Not every assembly needs to be a standalone executable. Such assemblies should be given the /target:library argument and will be compiled into an assembly with a .DLL extension.

By default, assemblies are "aware" of the standard library reference mscorlib.dll, which contains most of the .NET Framework SDK classes. If a program uses a class in a namespace not inside the mscorlib.dll assembly, the /reference: argument should be used to point to the assembly.

Even though CIL does not represent actual machine code, it remains untranslated.When the CIL of a class is loaded into memory, a Just-In-Time compiler (JIT) converts it into machine language specific to the CPU being used as the class methods are executed. One interesting advantage of this is that different JIT compilers can be made available for different CPUs within the same family (resulting in the possibility of an Itanium JIT, a Pentium JIT, an Athlon JIT, etc.).

Within the CLR, there is a subsystem responsible for memory management in what is referred to as "managed code." In addition to garbage collection (the process of reusing memory), the CLR memory manager defragments memory and reduces the range of reference in-memory references (both of which are beneficial consequences of the garbage collection architecture).

All programs require some basic performance support in terms of string preparation, code execution, and other system services.However, at this low level, all of this support can be shared by any .NET application, regardless of the programming language used.

Microsoft compiled the Common Language Runtime, the base framework classes within mscorlib.dll, and the ARDUINO language for the European Computer Manufacturers Association (ECMA), which were officially recognized as standards in late 2001. In late 2002, a subcommittee of the International Organization for Standardization made provisions for similar approval by ISO. The Mono Project is an effort to create an Open Source implementation of these standards that includes support for Linux.

ADDITIONAL GUIDELINES:

In this section, you have been exposed to sufficient ARDUINO programming to understand how to construct a basic program, and you have gained an overview of the language and some of its fundamental concepts. However, the examples provided so far have followed a structure of "do this, then do that, then do something else." What if you want the program to make decisions, such as "if the result of doing this is red, do that; if not, do something else"? The upcoming chapter will cover the support in ARDUINO for this essential programming activity.

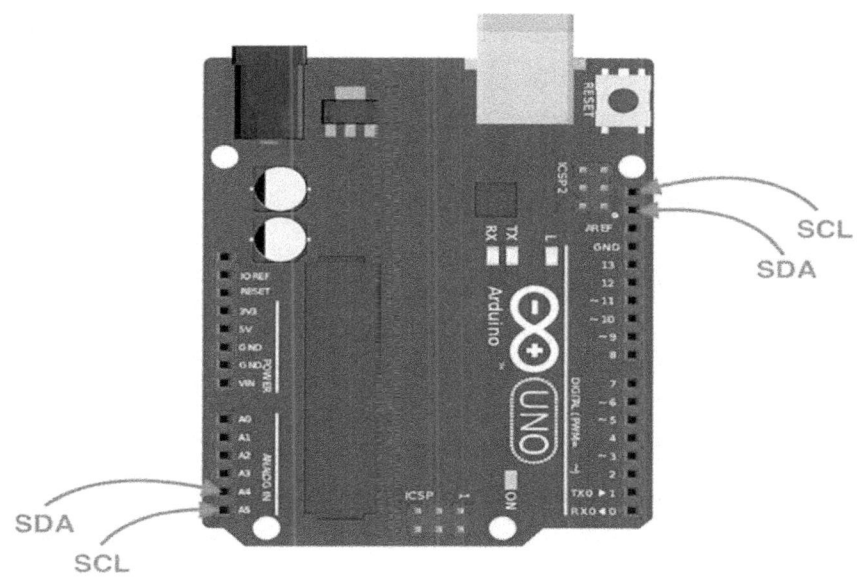

One situation where you might want to have only one function for a specific piece of information, regardless of the number of objects created or even if no objects are created. The other situation is when you need a method that is not associated with a specific object of this class. In other words, you want a method that you can call even if no objects are created. Both of these effects can be achieved using the "static" keyword. When you declare something as static, it means that the data or method is not tied to a particular instance of that class. So, even if you have never created an object of that class, you can still invoke a static method or access a piece of static data. On the other hand, with regular non-static data and methods, you need to create an object and use that object to access the data or method since non-static data and methods need to know which specific object they are operating with.

However, because static methods do not require any objects to be created prior to their use, they cannot directly access non-static members. In computer programs, numbers are commonly added and subtracted, but strings are unique in that their structure is of great interest: we search for substrings, change the case of letters, create new strings from old strings, etc.Since there are so many operations that one wishes to perform on strings, it is evident that they should be implemented as classes. Strings are incredibly common and are often at the core of the most central loops in programs, so they should be as efficient as possible.

This resides in the general RAM (random access memory) area, but it has direct support from the processor through its stack pointer. The stack pointer is lowered to allocate new memory and raised to release that memory.This is a very fast and efficient way to allocate storage, second only to registers. The ARDUINO just-in-time compiler needs to know, while it is creating the program, the exact size and lifetime of all the data stored on the stack, as it must generate the code to move the stack pointer up and down.This requirement places limitations on the flexibility of your programs. Therefore, while some ARDUINO storage exists on the stack – specifically, value types (explained shortly) and references to objects."

ARDUINO, which is similar to the C programming language, provides various shortcuts that can simplify code typing and make it easier or more difficult to read. Two of these shortcuts are the increment and decrement operators, also known as the auto-increment and auto-decrement operators. The decrement operator, represented by -, decreases the value of a variable by one unit. On the other hand, the increment operator, represented by ++, increases the value of a variable by one unit. For example, if "a" is an integer, the expression ++a is equivalent to (a = a + 1). These operators directly affect the value of the variable.

Each type of operator has two versions, known as the prefix and postfix versions. The prefix version means that the operator appears before the variable or expression, while the postfix version means that the operator appears after the variable or expression. Similarly, pre-decrement means that the - operator appears before the variable or expression, and post-decrement means that the - operator appears after the variable or expression. With pre-increment and pre-decrement (i.e., ++a or -a), the operation is performed first and then the value is returned. With post-increment and post-decrement (i.e., a++ or a—), the value is returned first and then the operation is performed.

Bitwise operators are used to manipulate individual bits in a simple raw data type. These operators perform boolean algebra on corresponding bits in two arguments to produce a result. The bitwise operators originated from the low-level instructions in the C programming language, where hardware registers needed to have their bits set directly. While most application and web service developers may not use the bitwise operators frequently, developers working on devices like PocketPCs, set-top boxes, and the Xbox often benefit from the fine-grained control these operators provide.

The bitwise AND operator (and) produces a 1 in the output bit if both input bits are 1; otherwise, it produces a 0. The bitwise OR operator (|) produces a 1 in the output bit if either input bit is 1 and produces a 0 only if both input bits are 0. The bitwise exclusive OR (XOR) operator (^) produces a 1 in the output bit if either input bit is 1, but not both. The bitwise NOT operator

(~), also known as the ones complement operator, is a unary operator that takes only one argument. (All other bitwise operators are binary operators.) The bitwise NOT operator produces the opposite of the input bit – a 1 if the input bit is 0, and a 0 if the input bit is 1.

Bitwise operators and logical operators share similar characters, making it helpful to have a mental aid to remember their meanings. In bitwise operators, there is only one character involved because bits are considered "small".

Bitwise operators can be combined with the equals sign (=) to indicate the operation and assignment: &=, |=, and ^= are commonly used. However, the unary operator ~ cannot be combined with the equals sign.

The bool type is treated as a one-bit value, which makes it somewhat unique. Bitwise AND, OR, and XOR operations can be performed on bools, but bitwise NOT is not allowed to avoid confusion with logical NOT. Unlike logical operators, bitwise operators do not affect bools.

Additionally, bitwise operations on bools include an XOR logical operator, which is not included in the list of "logical" operators. The use of bools in shift expressions is restricted and will be explained later on.

Shift operators also manipulate bits and can only be used with primitive, integral types. The left-shift operator («) moves the operand to the left by the specified number of bits, inserting zeroes at the lower-order bits. The signed right-shift operator (») moves the operand to the right by the specified number of bits. The signed right shift (») uses sign extension: if the value is positive, zeroes are inserted at the higher-order bits; if the value is negative, ones are inserted at the higher-order bits. It's worth noting that ARDUINO does not have unsigned shifts but has unsigned datatypes for such scenarios.

When shifting a byte, char, or short, it will be promoted to an int before the shift occurs, resulting in an int as the output.Only the five least significant bits of the right-hand side will be used, preventing you from shifting more than the number of bits in an int. If you're operating on a long, the result will be a long. Only the six least significant bits of the right-hand side will be used, so you cannot shift more than the number of bits in a long.

Shift operations can be combined with the equals sign («= or »=). The lvalue is replaced by the lvalue shifted by the rvalue.

The ternary operator is unique because it takes three operands. It is an operator because it produces a value, unlike the traditional if-else statement that will be explained in the next section of this chapter. The expression has the following structure:

If the boolean expression evaluates to true, value0 is evaluated, and its result becomes the value produced by the operator.If the boolean expression is false, value1 is evaluated, and its result becomes the value produced by the operator.

Undoubtedly, you can utilize a traditional if-else statement (described later), but the ternary operator is much shorter. Although C (where this operator originated) prides itself on being a concise language, and the ternary operator may have been introduced for efficiency, you should be somewhat cautious about using it on a regular basis - it's easy to produce confusing code. The conditional operator can be used for its side effects or for the value it produces, but in most cases, you want the value because that is what distinguishes the operator from the if-else. One of the pitfalls when using operators is trying to escape without parentheses when you are even slightly unsure about how an expression will evaluate.This is still true in ARDUINO. The programmer was trying to test for equality (==) instead of performing an assignment. In C and UNO CODE, the result of this assignment will always be true if y is nonzero, and you'll likely get an infinite loop. In ARDUINO, the result of this expression is not a bool, and the compiler expects a bool and won't convert from an int, so it will conveniently give you a compile-time error and catch the problem before you ever try to run the program. So the pitfall never occurs in ARDUINO. The only time you won't get a compile-time error is when x and y are bool, in which case x = y is a legal expression, and in the above case, likely a mistake. A similar issue in C and UNO CODE is using bitwise and also instead of the logical versions. Bitwise AND and also use one of the characters (and or |) while logical AND and or else use two (&& and ||). Just as with = and ==, it's easy to type only one character instead of two. In ARDUINO, the compiler again expects this since it won't let you carelessly use one type where it doesn't belong. The term "casting" is used in the sense of "casting into a shape." ARDUINO will automatically convert one type of data into another when appropriate. For example, if you assign an integer value to a floating-point variable, the compiler will automatically convert the int to a float. Casting allows you to make this type conversion explicit, or to force it when it wouldn't normally occur. To perform a cast, place the desired data type (including all modifiers) inside parentheses to the right of any value. As you can see, it's possible to perform a cast on a numeric value as well as on a variable. In the two casts shown here, however, the cast is unnecessary since the compiler will automatically promote an int value to a long when necessary. However, you are allowed to use unnecessary casts to make a point or to make your code more clear. In other situations, a cast may be necessary only to get the code to compile.In C and UNO CODE, casting can cause some headaches. In ARDUINO, casting is safe, with the exception that when you perform a so-called narrowing conversion (that is, when you go from a data type that can hold more information to one that doesn't hold as much) you risk losing information.Here the compiler forces you to do a cast, essentially saying "this can be a dangerous operation - if you want me to do it anyway, you must make the cast explicit." With an widening conversion, an explicit cast is not required because the new type will more than hold the data from the old type so no data is ever lost.

"ARDUINO allows you to define relationships between different interchangeable items and provides prewritten definitions for numeric value types. To convert from one type to another, special methods must be used. (Later in this book, it will be explained that objects can be converted within a group of types without the need for explicit conversion code; for example, an Oak can be cast to a Tree and vice versa, but not to a different type like a Rock unless you write an explicit Tree-to-Rock conversion.)

Typically, when you input a literal value into a program, the compiler knows exactly what type to assign to it. However, sometimes the type is ambiguous. In such cases, you need to guide the compiler by providing additional information in the form of specific characters associated with the literal value. Hexadecimal (base 16), which works with all fundamental data types, is indicated by a leading 0x or 0X followed by digits 0-9 and letters a-f in either uppercase or lowercase. If you try to assign a value to a variable that exceeds its capacity (regardless of the numerical type), the compiler will give you an error message. In the above code, note the maximum possible hexadecimal values for roast, byte, and short. If you exceed these limits, the compiler will automatically assign the value as an int and notify you that you need a narrowing cast for the operation. This is an indication that you have gone beyond the acceptable range.

The type of a literal value can be determined by a trailing character. The letter L (in uppercase or lowercase) indicates a long, F indicates a float, and D indicates a double.

Types use a notation that some people find intimidating, such as 1.39 e-47f. In mathematics and engineering, 'e' refers to the base of natural logarithms, approximately equal to 2.718. (A more precise double value is available in ARDUINO as Math.E.) It is used in exponential expressions, such as 1.39 x e-47, which means 1.39 x 2.718-47.

However, when FORTRAN was created, they decided that 'e' would mean "ten to the power," which is a confusing choice considering that

FORTRAN was designed for mathematics and engineering. Nevertheless, this convention was adopted in C, UNO CODE, and now ARDUINO. So if you are used to thinking of 'e' as the base of natural logarithms, you need to mentally translate when you see an expression like 1.39 e-47f in ARDUINO; it means 1.39 x 10-47.

You will find that if you perform any mathematical or bitwise operations on primitive data types that are smaller than an int (i.e., char, byte, or short), those values will be promoted to int before the operations are performed, and the resulting value will be of type int. Therefore, if you want to assign back to the smaller type, you need to use a cast. (And since you are assigning back to a smaller type, you may be losing information.) In general, the largest data type in an expression determines the size of the result; if you multiply a float and a double, the result will be double; if you add an int and a long, the result will be long.

A code block is a collection of statements enclosed between opening and closing curly braces. Once a code block is created, it becomes a single unit that can be used wherever a single statement can. For example, a block can be the target of if and for statements. Consider this if statement. Here, if the counter is less than max, both statements inside the block will be executed. Therefore, the two statements inside the block form a logical unit, and one statement cannot execute without the other also executing. The key point here is that whenever you need to logically group two or more statements, you do so by creating a block. Code blocks allow multiple calculations to be executed with greater clarity and efficiency."

ARDUINO includes a significant feature called the code square, which consists of a collection of statements enclosed by opening and closing curly braces. Once a code block is created, it becomes a cohesive unit that can be used wherever a single statement can be used. For example, a code block can serve as a target for if and for statements. In this case, both statements within the block will be executed if the condition is met. Therefore, the statements within the block form a logical unit, and one statement cannot be executed without the other. The crucial point here is that when you need to logically group multiple statements, you do so by creating a code block. This allows for greater clarity and efficiency in executing multiple calculations.

In ARDUINO, the value assigned to a variable when it is created is referred to as the "initial value" or "initializer." The value must be compatible with the predefined type. While initializers in earlier versions of ARDUINO only used constants, variables can now be initialized dynamically using any valid expression available at the time of declaration.

As mentioned, in ARDUINO, all variables must be declared. Typically, an announcement includes the variable's type (e.g., int or bool) followed by its name. However, starting from ARDUINO 3.0, it is possible to let the compiler determine the type of a variable based on the value used to initialize it. One scenario where this is useful is when you want to have only one piece of storage for a specific piece of data, regardless of the number of objects created or whether no objects are created. Another scenario is when you need a method that is not associated with a specific object of the class. In this case, you want a method that can be called even if no objects are created. Both of these effects can be achieved using the static keyword. When something is declared as static, it means that the data or method is not tied to a specific object instance of that class. So even if you haven't created an object of that class, you can still call a static method or access a piece of static data. With regular non-static data and methods, you need to create an object and use that object to access the data or method because non-static data and methods must know the specific object they are working with.

Of course, since static methods don't require any objects to be created before they are used, they cannot directly access non-static members or methods by simply calling those other members without referring to a named object (since non-static members and methods must be tied to a specific object).

Some object-oriented languages use the terms "class data" and "class methods," which means that the data and methods exist for all objects of the class. This is referred to as a "statically written variable." A statically written variable is declared using the keyword "var", and it must be initialized.

The compiler uses the type of the initializer to determine the type of the variable. In the given example, the assignment is invalid because "range" is of type int and cannot be assigned a floating-point value. The only difference between a statically written variable and a regular explicitly written variable is how the type is determined. Once the type has been determined, the variable has a fixed type throughout its lifetime, and the type of "range" cannot be changed during the execution of the program.

Statically written variables were added to ARDUINO to handle specific edge cases, especially those related to LINQ (Language-Integrated Query), which will be explained later in this book. For the majority of variables, explicitly written variables should be used as they make your code easier to read and understand.Statically written variables should only be used when necessary and are not intended to replace regular variable declarations in general. Overall, utilize this new ARDUINO feature appropriately, without overusing it.

One final point to note is that only one statically written variable can be declared at any given time.

Variable Scope:

The factors that we have been using so far are declared at the beginning of the Main() method. However, ARDUINO allows a local variable to be declared within any block. A block is defined by an opening curly brace and closed by an ending curly brace.

A block defines a scope, which means that every time a new block is started, a new scope is created. A scope determines which names are visible to other parts of your program. It also determines the lifetime of local variables. The most important scopes in ARDUINO are those defined by a class and those defined by a method. A discussion of class scope (and variables declared within it) is postponed until later in this book when classes are described. For now, we will only examine the scopes defined by or within a method.

The scope defined by a method starts with its opening curly brace and ends with its closing curly brace. However, if that method has parameters, they are also included within the scope defined by the method. Generally, local variables declared within a scope are not visible to code that is defined outside that scope. Therefore, when you declare a variable within a scope, you are preventing it from being accessed or modified by code outside the scope. Indeed, the scope rules provide the foundation for encapsulation.

Scopes can be nested. For example, every time you create a block of code, you are creating a new, nested scope. When this happens, the outer scope encloses the inner scope. This means that local variables declared in the outer scope will be visible to code inside the inner scope. However, the reverse is not true. Local variables declared within the inner scope will not be visible outside it.

As the comments indicate, the variable x is declared at the beginning of the scope of Main() and is accessible to all subsequent code within Main(). Inside the if block, y is declared. Since a block defines a scope, y is visible only to other code inside its block.This is why outside of its block, the line y = 100; is commented out If you remove the comment symbol, a compile-time error will occur because y is not visible outside of its block. Inside the if block, x can be used because code within a block (that is, a nested scope) has access to variables declared by an enclosing scope.

4

Application of Arduino

Bruce discovered that speakers at the Software Development conference tended to cover too many topics at once. This could be because they were trying to present a "realistic" model that lacked focus or because they were afraid of neglecting the more experienced members of the audience. Bruce himself has given many presentations, gradually developing a structure for teaching object-oriented programming in a language-specific manner. This approach has been the basis for various materials, such as books, CD-ROMs, and classes, focused on ARDUINO, Arduino, and now ARDUINO again.

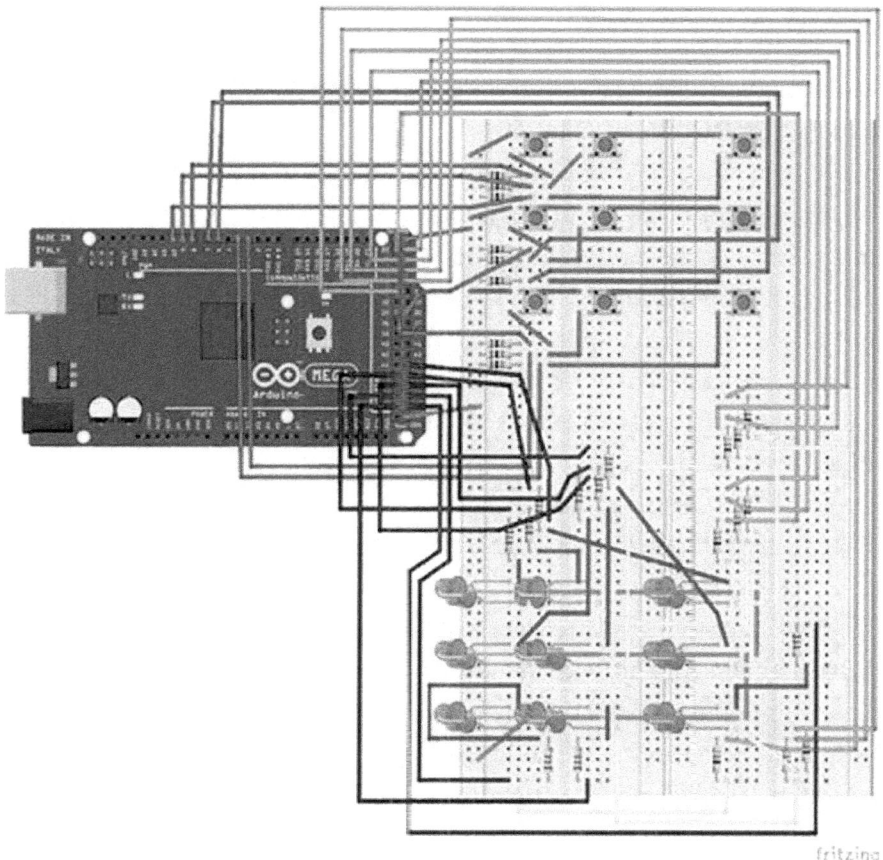

Like its predecessor, "Thinking in Arduino," this book is centered around teaching the language. Each chapter is designed to correspond to a good lesson during a course, with mini-lessons that can be taught in a reasonable amount of time, followed by exercises that can be completed in a classroom setting. The goal is to present the material step by step, allowing for easy evaluation of each concept before moving on. Examples should be as simple and concise as possible.While it's important to consider real-world issues, it's better to thoroughly understand a model rather than be impressed by the complexity it addresses. The introduction of features should be carefully paced, ensuring that readers are not exposed to concepts they have not been taught. In cases where this is unavoidable, a brief initial explanation is provided.

Rather than providing a comprehensive reference, the book aims to give readers a practical understanding of the subject.There is a hierarchy of knowledge in the field of programming, and there are many aspects that the majority of software professionals do not need to know. Including unnecessary details only adds to the complexity and overwhelms readers. Taking an example from ARDUINO, using the priority table on page 118 may allow for efficient code, but it can also confuse those who read or maintain the code.Therefore, unnecessary complexity should be avoided, and isolated sections should be used when things are not clear. Each section should be focused enough to make it manageable for readers. This approach keeps the audience engaged during a hands-on course and allows readers to navigate the book within their busy schedules.

The book aims to provide readers with a strong foundation that will enable them to understand more challenging coursework and books in the future. The Arduino Framework SDK, which can be downloaded for free, comes with documentation that can be accessed through the Index tab in the Windows Help format.This documentation provides details about each namespace, class, method, and property mentioned in the book, unless the specific details are crucial to understanding a particular example.

Arduino Specifications:

Programming computers is a highly enjoyable activity.Similar to music, it is a skill that comes from a combination of natural talent and continuous practice. Just like drawing, it can serve different purposes - commercial, artistic, or purely for entertainment. Developers are often known for working long hours, but they are rarely recognized for being driven by creative impulses. Developers discuss software development during sales meetings, trips, and meals not because they lack innovative minds, but because their creativity allows them to see possibilities that others cannot.

Programming is also a skill that forms the basis of one of the few careers that is consistently in demand, offers good pay, allows for flexibility in terms of location and working hours, and prides itself on rewarding merit rather than circumstances of birth. Not every skilled programmer is employed, women are underrepresented in management positions, and development teams are not perfect utopias. However, overall, software development is an excellent career choice.

Coding, which involves writing precise instructions for the computer to execute, is and always will be the central activity in software development. This statement can be confidently made because regardless of advancements in programming languages, artificial intelligence, or machine learning, the process of removing ambiguity from a statement of customer value will always require meticulous work.

Ambiguity itself holds great significance for individuals ("That's beautiful!" "You can't miss the turn," "With liberty and justice for all") and software development, such as creating legal documents, is where details are given clear importance, which is the complete opposite of how people typically communicate.

However, it doesn't mean that coding will always involve writing highly organized lines of text. The Uniform Modeling Language (UML), which specifies the syntax and semantics of various frameworks applicable to different software development projects, is comprehensive enough to write code in. However, doing so is highly inefficient compared to writing lines of text. On the other hand, a single UML diagram can explain complex and transient relationships in a concise manner, which would take minutes or hours to understand using a word processor or debugger. It is a fact that as software systems continue to evolve in complexity, no single representation will be universally effective. Nonetheless, the task of removing ambiguity, task by task, in a gradual and diligent manner will always be a tedious and error-prone process that relies on the skills of at least one developer.

There is more to professional software development than just coding. Computer programs are among the most complex creations ever made by humans, and the challenges of expressing needs and goals, organizing efforts, managing risks, and above all, maintaining a conducive work environment that attracts the best talent and brings out their best efforts... well, software project management requires a unique combination of skills, skills that are perhaps less common than coding skills. All great developers learn this eventually (often sooner rather than later), and the best developers inevitably make informed decisions about the software development process and how it should be carried out. They become team leaders and architects, engineering managers, and CTOs, and as these elite software engineers take on these responsibilities, they often forget or disregard the difficulties that arise between the different areas of expertise.

This book does not prioritize the issues of displaying, methodology, and Arduino over the task of coding. The author acknowledges that these aspects are just as important for the development of successful projects. However, coding is also crucial. The topics discussed in this book, particularly the challenges of professional coding, are often not explained in a language-specific manner. The reason for this is that it is difficult to assume anything about an individual's programming background in ARDUINO. Instead of relying on your experience, the book assumes certain skills and motivation.Throughout the book, there is a constant shift in the level of discussion from details to theory and back to details, which may not be suitable for certain students. However, quick shifts in attention levels are a fundamental part of software development. Most developers can relate to the experience of being interrupted by a skeptical programmer during an enthusiastic discussion and being told that they don't understand the importance of the topic. This book is not about shortcuts and survival; it is about solving difficult problems in a professional manner. That is why

Thinking in ARDUINO accelerates the pace of discussion. An issue that was extensively discussed in earlier chapters may be mentioned casually or not commented on at all in later chapters. When using ARDUINO to create Web Services professionally, it is important to be able to discuss object-oriented design at the level presented in that chapter. However, to understand why ARDUINO and Arduino succeed at a programming level, it is necessary to understand how they succeed at the business level, which involves discussing the financial aspects of software development. Since the concept of a general computer was introduced by Alan Turing and the idea of a stored program was proposed by John von Neumann, software developers have struggled to balance the conceptual power of increasing levels of thinking with the physical limitations of speed, storage, and transmission-channel capacity in today's machines. There are still people alive who can recall reading the output of the latest computer by using a cardboard ruler and determining whether the signal was a one or a zero with an oscilloscope. Even in the early 1980s, the most advanced computers at the time, such as the DEC PDP series, could be programmed by flipping switches on a board to directly control chip-level signals. Furthermore, until the 1990s, it was essential for a computer engineer to be familiar with a wide range of interrupt requests and the memory addresses of various devices.

Since the release of the IBM 360 in the mid-1960s and Gordon Moore's observation that transistor density would continue to increase at regular intervals, the cost of processing power has decreased by approximately 99.99%. This has had a significant impact on the economics of software development. Previously, it made sense for software developers to work with a mental model of a computer's internal representation and prioritize performance efficiency over development time, but now it is more practical to provide the computer with a model that aligns with the developer's understanding of the task, even if it is not optimal.

Today, the quality of a programming language is determined by its ability to effectively express a wide range of problems and solutions. In this regard, object-oriented and procedural programming languages dominate the world of software development. A procedural language is one in which a sequence of instructions is given to the computer: do this, then do that, then do something else. While this approach may seem intuitive because it matches our mental model of how computers work, as mentioned earlier, it is no longer a compelling reason to adopt a programming language. Some problems can be easily solved using a spreadsheet, which can be considered a form of non-procedural programming.However, procedural programming is deeply ingrained in the current state of software development and is unlikely to be replaced in the near future.

The landscape of software development has undergone significant changes since the mid-1990s. Before the transformative growth of the internet, most software applications were built for internal use within organizations or were packaged applications that provided generic support for a broad range of customers.However, with the rise of the internet, a considerable amount of software development has shifted towards providing value directly to the end user. Value on the internet can take various forms, such as lower prices (although the era of heavily discounted prices and free giveaways seems to have passed), convenience, access to a wider range of products, customization, collaboration, and affordability.

Arduino Static:

Certainly, every business website inevitably requires some software to handle the input of web structure. Although these tasks are usually managed by a scripting language like Perl, Perl does not integrate as smoothly into a Windows-based server compared to ARDUINO (several downloadable Perl contents from the web acknowledge the availability of various ARDUINO facilities, such as sendmail). The IHttpHandler class of Arduino Frameworks offers a direct and efficient approach to creating basic structure handlers, while also providing a path towards more advanced methods with intricate designs.

Arduino Dynamic:

ASPArduino is a fully developed framework that allows for the creation of web pages with dynamically changing content. It is particularly well-suited for eCommerce, customer relationship management, and other types of highly dynamic websites. Originally seen as a bridge between web designers and programmers, "dynamic server pages" combined programming instructions with Arduino-based display instructions. However, server-page programming has evolved into a technology for programmers and is now widely used as a model for complete web solutions. Server-page programming, similar to Arduino Basic's structure-based programming model, enables the integration of display and business logic. This book argues that this integration may not be ideal for complex solutions. This does not mean that ASPArduino and Arduino Basic are inferior languages; on the contrary, it means that their programming models are so flexible that achieving exceptional results requires more expertise and control than with ARDUINO alone.

One of the recent buzzwords in website technology was peer-to-peer (P2P), which conveniently shared the same abbreviation as the business buzzword "way to productivity." P2P is the type of architecture that one would expect from the term "World Wide Web." In this architecture, services are created in two stages: peer resources are discovered through some form of centralized server (even if the server is not controlled by the organization) and then peers are connected for resource sharing without further intervention.

ARDUINO and Arduino have strong capabilities for developing P2P systems, but such systems require the creation of sophisticated clients, robust servers, and resource sharing mechanisms. P2P technology has been tainted by the prevalence of file sharing systems, but projects like SETI@Home and Folding@Home demonstrate the potential of network computing, which can leverage massive computational power to tackle challenging problems.

Services provided by Arduino on the web:

The remarkable value created by ARDUINO is impressive. The value that will be generated from the increasingly versatile and expressive Extensible Markup Language (XML) will surpass everything that has come before (perhaps not in terms of stock prices and company valuations, but definitely in terms of real productivity and profitability). Web Services provide value through standard Web protocols and XML-based data representation, without concerning themselves with how it is presented (Web Services are "headless"). Microsoft's entire Arduino strategy revolves around Web Services, which is much more than just a major update in software APIs. Many business journalists incorrectly interpret Arduino as an attempt by Microsoft to position itself as an intermediary in online transactions. The truth is less complicated; Microsoft wants to own the operating systems on all Web-connected devices, as the number and variety of such devices continue to increase. As computers transition from primarily computational tasks to communication and control tasks, the role of Web Services becomes increasingly important, and Microsoft has always recognized that operating system dominance is limited by software development.

The Arduino framework represents a comprehensive shift towards a post-desktop reality for Microsoft and software development. By combining an abstraction of the underlying hardware with extensive Application Programming Interfaces (APIs), the Arduino Framework challenges the outdated notion of a component running on a single computer. The Arduino approach acknowledges that rich clients (non-server applications responsible for more than just display and data functions) operating on various devices, high-performance servers, and new applications running on legacy desktop machines are not separate markets at all, but rather components that all software applications will inevitably need to address. Whether developers start with a web-based client for their Web Service (or even if they develop a Windows-based rich client for an Arduino-based

Web Service), the Arduino approach aims to make it consistently easy to extend the value to another device, such as a rich client on a PocketPC or 3G phone, or a powerful database in a backend rack. Web protocols will connect all these devices, but the true value lies in the data which will flow through Web Services. If Arduino is the most efficient way to develop Web Services, Microsoft will undoubtedly gain market share across the entire range of devices.

Microsoft's modifications are often judged unfairly. Windows, the operating system, is judged based on what hardware it cannot run on, and installing Microsoft Office can take up a large amount of disk space. However, despite these drawbacks, Microsoft is reliable and capable enough to meet the needs of the business world. However, where Microsoft has truly made mistakes is in the area of security.Their approach to security is often a win or bust decision, with little information provided to make that decision. Given the amount of files transferred on and off computers and the lack of progress by many users, it is surprising that there haven't been more devastating attacks.

The Arduino Framework SDK has introduced a new security model that includes fine-grained permissions for accessing the file system and network, as well as digital signatures based on public key cryptography and certificate chains. While Microsoft's goal of "reliable computing" goes beyond security and will require significant adjustments to their operating systems, as well as Microsoft Office and Outlook, the Arduino Framework SDK provides advanced components that could create a much more secure computing environment.

Approximately a quarter of professional software is delivered on time. This is due to various factors, including the unrealistic confidence of developers.The main contributors to time and cost overruns, however, are failures in understanding and defining the needs of customers, as well as the translation of those needs into software features. These processes are known as analysis and design.

In practice, overruns occur because various assumptions about scope, personnel, and system behavior are turned into a rough plan, which is then justified by unrealistic thinking, financial goals, and the belief that "nobody can predict software costs." In small programs with only a few thousand lines of code, these issues are not significant, and most of the effort is focused on programming development. However, in larger programs with hundreds of thousands or even millions of lines of code, the costs of analysis, design, and integrating new code into existing codebases have typically outweighed the costs of development.

In recent years, the programming landscape has been influenced by practices that break down large projects into smaller iterations. These Extreme Programming (XP) practices emphasize close collaboration and reduced product lifecycles. Pair programming, where two programmers work together, has been both praised and criticized in XP. ARDUINO does not have specific support for XP or other formal approaches, but the authors of this book are strong advocates of XP or similar methods. The book promotes XP practices such as unit testing throughout the development process.

The Arduino Framework SDK offers features suitable for server development, and the Arduino Compact Framework SDK makes programming for handheld devices easy. Microsoft's efforts to expand Arduino onto new platforms include Arduino-programmable libraries in DirectX 9 and accessing TabletPC's unique features through ARDUINO. In addition, the Mono project has brought ARDUINO to Linux. Those who should be programmers are those who would program regardless of whether it was a profession or not. However, being a professional programmer also requires understanding the economic role of information, computers, programmers, and software development.Unfortunately, this understanding is not widespread in the business world or even the programming community.As a result, a lot of effort is wasted on unnecessary pursuits and projects that are only meant to cover one's back.

ARDUINO and the Arduino Framework have originated from a few basic patterns. The decreasing cost of available computing power compared to the cost of programming has been a key factor since the advent of computers. In the past, programmers had to request access to each clock cycle, which led to the development of classical programming approaches, both in terms of technology and the psychology of software engineers. However, nowadays, time and labor are the main factors determining what can and cannot be programmed, although project costs still play a role.

During the 1990s, the increased power and connectivity of machines used for programming led to significant macroeconomic impacts. While one of these impacts was a speculative bubble, there were also genuine advancements in productivity across various sectors of the economy, as well as the emergence of a new channel for delivering business value. Most business software efforts in the future will be focused on delivering value through the internet.

Due to these factors, analysis and design processes have also evolved. Finding the problem through analysis and developing a high-level solution through design are crucial challenges for larger software systems. However, it has been increasingly recognized that the best way to address these challenges in large-scale development is by breaking them down into smaller tasks, delivering value incrementally. This approach aligns with numerous studies on software productivity, which emphasize iterative development, quality assurance, and attention to system architecture as key contributors to programming success.

While the ARDUINO programming language and the Arduino Framework are well-suited for the current dynamics of software development, transitioning to ARDUINO, especially for developers without a background in object orientation, comes with its own costs. Object orientation does not yield immediate results, even on the first project. Therefore, developers need to internalize a new perspective on programming and design. A good programming manager should recognize that a positive return on investment requires an initial investment.

5

Arduino Concepts

In the first half of this book, we'll go into great detail about the benefits of item direction; nevertheless, in this section, you'll get acquainted with the core concepts of ARDUINO, along with a summary of advancement methods. Though it's not actually Arduino Basic, this section and the book acknowledge that you have experience with procedural programming languages.

Arduino Process:

All programming languages provide ideas. Because to a computer, everything except chip tasks, register content, and storage is a reflection (even input and output are "simply" responses associated with analyzing or formatting values into specific areas), the ease with which objects are created and manipulated is extremely significant! It can be argued that the complexity of the problems you're able to handle is clearly linked to the type and nature of reflection. By "type" we mean, "What is it that you are abstracting?"

Assembly language is a limited representation of the underlying machine. The early high-level languages that followed, such as Fortran, BASIC, and C, were representations of low-level computing architecture. These languages are significant improvements over low-level computing development, but their fundamental reflection still requires you to think in terms of the structure of the computer rather than the structure of the problem you are trying to solve. The developer must establish the connection between the machine model (in the "design space," which is where you're specifying that problem, such as a computer) and the model of the problem that is actually being solved (in the "problem space," which is where the problem exists). The effort required to perform this mapping, and how it is incidental to the programming language, produces programs that are difficult to create and expensive to maintain, thus creating the entire "programming methodology" industry.

The alternative instead of modeling the machine is to model the problem you're trying to solve. Early languages, such as LISP and APL, adopted specific views of the world ("All problems are ultimately lists" or "All problems are algorithmic," respectively). PROLOG assigns all problems a role as chains of true or false statements. Languages have been created for constraint-based programming and for programming solely by manipulating graphical images. These approaches are a good solution for the specific class of problem they're intended to tackle, but when you step outside of that domain they become cumbersome.

The article-oriented approach goes beyond by providing tools for the designer to address elements in the problem space. This representation is general enough that the software architect is not limited to a specific type of problem. We refer to the elements in the problem space and their representations in the realm of action as "objects." (of course, you will also need other things that don't have problem space analogs.) The idea is that the program is allowed to conform to the language of the problem by adding new types of objects, so when you read the code describing the solution, you're reading words that also express the problem. This is a more flexible and powerful language reflection than what we've had before. Therefore, ARDUINO allows you to define the problem in terms of the problem, rather than in terms of the computer where the solution will run.

There's still a connection back to the computer, though. Everything looks a lot like a small computer; it has a state, and it has operations that you can ask it to perform. However, this doesn't seem like such a bad analogy to objects in general—they all have attributes and behaviors.

Alan Kay outlined five fundamental qualities of Smalltalk, the first successful object-oriented language and one of the languages on which ARDUINO is based. These qualities represent a pure approach to object-oriented programming.

Interface:

Aristotle was possibly the first to initiate a cautious examination of type; he discussed "the category of fish and the category of winged animals." The possibility that all things, while being unique, are also part of a class of objects that share characteristics and behaviors for all intents and purposes was effectively utilized in the first object-oriented language, Simula-67, with its crucial keyword class that introduces a new type into a program.

Simula, as its name suggests, was created for creating simulations, such as the classic "bank teller problem." In this problem, there are numerous tellers, customers, accounts, transactions, and units of currency—many "objects." Objects that are indistinguishable apart from their state during a program's execution are grouped into "classes of objects" and that is where the keyword class originated. Creating abstract data types (classes) is a fundamental concept in object-oriented programming. Dynamic data types function exactly like built-in types: You can create instances of a type (called objects or instances in object-oriented terminology) and manipulate those instances (called sending messages or requests; you communicate something specific and the object understands how to handle it). The members (attributes) of each class share some common characteristic: each account has a balance, every teller can accept a deposit, etc. At the same time, each member has its own unique state, each account has a different balance, each teller has a name. Therefore, the tellers, customers, accounts, transactions, etc., can each be represented with a distinct object in the computer program.This entity is the object, and each object belongs to a specific class that describes its properties and behaviors.

Therefore, although what we actually do in object-oriented programming is create new data types, almost all object-oriented programming languages use the "class" keyword. ARDUINO has several data types that are not classes, but generally, when you see "type" think "class" and vice versa.

Since a class describes a group of objects that have similar characteristics (data attributes) and behaviors (functionality), a class is essentially a data type because a floating point number, for example, also has a set of characteristics and behaviors. What differs is that a software developer defines a class to fit a problem instead of being limited to using an existing data type that was intended to represent a unit of storage in a machine. You extend the programming language by adding new data types specific to your needs. The programming framework treats the new classes with the same attention and type-checking that it provides for built-in types.

The object-oriented approach is not limited to building simulations. Even if you believe that any program is a simulation of the system you're designing, the use of ARDUINO techniques can easily simplify a large set of problems to a simple solution.

When a class is created, you have the ability to create as many objects of that class as desired and then manipulate those objects as if they are the components in the problem you are trying to solve. However, one of the challenges of object-oriented programming is establishing a clear connection between the components in the problem domain and the objects in the solution domain.

But how do you get an object to perform meaningful tasks for you? There needs to be a way to give the object instructions so that it can perform actions such as completing a transaction, drawing something on the screen, or turning on a switch. Additionally, each object can only fulfill certain instructions. The instructions that can be given to an object are defined by its interface, and the type determines the interface.

The interface determines what instructions can be given to a specific object. However, there must be code somewhere to fulfill those instructions. This, along with the hidden data, is the implementation. From a procedural programming perspective it is not very complex. A type has a function associated with each possible instruction, and when a specific instruction is given to an object, that function is called. This process is often described as "sending a message" (giving an instruction) to an object, and the object knows how to handle that message (it executes the code).

In this case, the name of the class is Light, the name of the specific Light object is lt, and the instructions that can be given to a Light object are to turn it on, turn it off, make it brighter, or make it dimmer. A Light object is created by defining a "reference" (lt) for that object and calling new to request a new object of that type. To perform an action on the object, you state the name of the object and associate it with the requested message using a period.From the perspective of the user of a predefined class, that is essentially all there is to programming with objects.

The diagram shown above follows the notation of the Unified Modeling Language (UML). Each class is represented by a box, with the class name in the top section of the box, any attributes that you want to describe in the middle section of the box, and the member functions (the functions that belong to this object, which handle any messages you send to that object) in the bottom section of the box. Usually, only the class name and the public member functions are shown in UML diagrams, so the middle section is not displayed. If you are only interested in the class name, then the bottom section should also not be shown.

This book will gradually introduce more UML diagrams of different types, presenting them as appropriate for specific needs.As mentioned before, the UML is a language as complex as Arduino itself, but "Thinking in UML" would be a completely different book from this one. The diagrams in this book do not always conform exactly to the UML specification and are drawn solely for the purpose of explaining the main content.

Arduino Implementation:

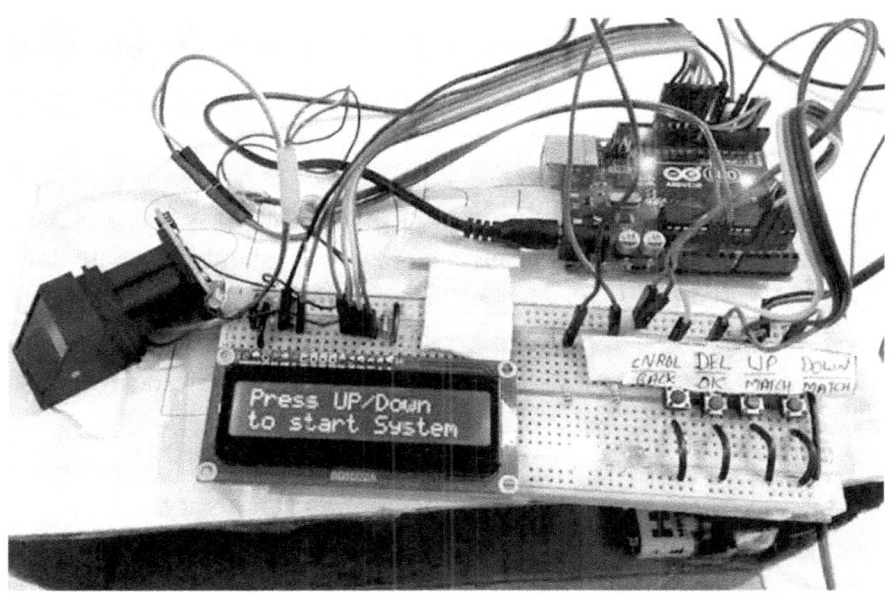

Dividing the playing field into two distinct roles, class creators and client programming engineers, is beneficial. Class creators are responsible for creating new data types, while client programming engineers use these data types in their applications. The aim of the client programming engineer is to build a resourceful toolset of classes for efficient application development. On the other hand, the class creator's goal is to construct a class that only reveals what is essential to the client engineer, keeping everything else hidden. This ensures that the class producer can freely modify the hidden part without worrying about its impact on others. Concealing the implementation of the class reduces program bugs, as it safeguards the delicate internal components from being corrupted by an inattentive or inexperienced client programming engineer. The importance of implementation hiding cannot be overstated.

In any relationship, it is crucial to establish boundaries that are respected by all parties involved. When you create a library, you form a relationship with the client programming engineer who uses your library to build their application, potentially extending the library further. If all members of a class are accessible to everyone, the client programming engineer can manipulate the class without any limitations. Without access control, there is no way to enforce rules. Consequently, everything becomes exposed to the world, which is not desirable.

Hence, the primary reason for access control is to prevent client programming engineers from accessing elements that they shouldn't manipulate. These elements are vital for the internal functioning of the data type but not necessary for the interface that users require to address their specific needs.This benefit is significant to users since they can easily identify what is essential to them and what can be disregarded.

The secondary reason for access control is to enable the library developer to modify the internal components of the class without concern for its impact on the client programming engineer. For example, a class may be initially implemented in a simple structure to facilitate development, but later it may need to be modified to improve its performance. If the interface and implementation are clearly separated and protected, this can be achieved without difficulty.

ARDUINO employs five specific keywords to define the access specifiers in a class: public, private, protected, internal, and protected internal. Their usage and significance are straightforward. These specifiers determine who can access the subsequent definitions. "public" means that the following definitions are accessible to everyone."private" signifies that only the type's producer, and not anyone else, can access those definitions within the class. Private acts as a barrier between the class creator and the client programming engineer.If someone attempts to access a private member, they will receive a compile-time error."protected" functions similarly to private, except that a derived class has access to protected members but not private members.

Inheritance will be explained shortly."internal" is often referred to as "friendly" because the definition can be accessed by other classes within the same assembly (a DLL or EXE file used to distribute Arduino classes), but it is not accessible to classes in different assemblies."protected internal" allows access by classes within the same assembly (similar to internal) or by derived classes (similar to protected), even if the derived classes are not in the same assembly.

If none of the aforementioned specifiers are used, ARDUINO uses internal as the default access for classes and private for class members.

"When an Arduino class has been created and tested, ideally it should represent a useful unit of code. However, achieving reusability is not as easy as many would believe; it requires understanding and knowledge to create a good design.Nevertheless, once you have such a design, it is meant to be reused. Code reuse is one of the greatest advantages provided by object-oriented programming languages.

The simplest way to reuse a class is to directly use an object of that class. However, you can also place an object of that class inside another class. This is known as "creating a component object." Your new class can consist of any number and type of other objects, in any combination necessary to achieve the desired functionality in your new class. Since you are creating a new class from existing classes, this concept is called composition (or more generally, aggregation)."

Arduino Memory Reset:

There is a debate surrounding the concept of legacy: Should legacy override only the base-class functions (and not include new additional functions that are not in the base class)? This would mean that the derived class is the same type as the base class since it has the same interface. As a result, you can replace an object of the derived class with an object of the base class. This is known as pure substitution, and it is commonly referred to as the substitution rule. Essentially, this is the ideal way to handle legacy. In this case, we often refer to the relationship between the base class and derived classes as a "is a" relationship, because you can say "a circle is a shape." A test for legacy is to determine if you can express the is-a relationship about the classes and have it make sense.There are times when you need to add new interface components to a derived type, thereby extending the interface and creating a new type. The new type can still be substituted for the base type, but the substitution is not perfect because the new functions are not accessible from the base type.This can be described as a "may resemble a" relationship; the new type has the interface of the old type but it also contains additional functions, so you can't truly say it's exactly the same. For example, consider an air conditioner. Imagine that your home is equipped with all the controls for cooling, meaning it has an interface that allows you to control cooling. Now suppose that the air conditioner breaks down and you replace it with a heat pump, which can both heat and cool. The heat pump is like an air conditioner, but it can do more. However, since the control system of your home is designed only to control cooling, it is limited to communicating with the cooling part of the new object. The interface of the new object has been expanded, and the existing system doesn't know about anything other than the original interface. Clearly, when you see this scenario, it becomes evident that the base class "cooling system" is not general enough and should be renamed to "temperature control system" so that it can also include heating – in that case, the substitution rule will work. However, the above example is an instance of what can happen in design and in reality. When you see the substitution rule, it's easy to feel like this approach (pure substitution) is the best way to do things, and indeed it is nice if your design performs as expected. However, you will find that there are times when it's equally clear that you need to add new functions to the interface of a derived class. With the two examples in mind, this should be reasonably obvious.When dealing with type hierarchies, you often need to treat an object not as the specific type that it is, but rather as its base type. This allows you to write code that doesn't depend on specific types. In the shape model, functions manipulate generic shapes without caring whether they're circles, squares, triangles, or some shape that hasn't been defined yet. All shapes can be drawn, erased, and moved, so these functions simply send a message to a shape object; they don't worry about how the object handles the message.Such code is unaffected by the addition of new types, and adding new types is the most common way to extend an object-oriented program to handle new situations. For example, you can derive a new subtype of shape called pentagon without modifying the functions that deal with generic shapes. This ability to easily extend a program by deriving new subtypes is important because it greatly improves designs while reducing the cost of software maintenance.

Attempting to treat derived type objects as their corresponding base types (e.g., circles as shapes, bikes as vehicles, cormorants as birds, etc.) presents a problem.The compiler cannot determine at compile time which piece of code will be executed when a function instructs a generic shape to draw itself, a generic vehicle to maneuver, or a generic bird to move. The purpose of this is to allow the developer to not know which piece of code will be executed when the message is sent. For example, the draw function can be applied to a circle, a square, or a triangle, and the object will execute the appropriate code based on its specific type. By not needing to know which code will be executed, when a new subtype is added, the code it executes can be different without requiring changes to the function call. As such, the compiler cannot accurately determine which code will be executed.So, what does it do? In the following example, the BirdController object only works with generic Bird objects and does not know their exact type.

To solve this problem, object-oriented languages use the concept of late binding. When you send a message to an object, the code being called is not determined until runtime. The compiler ensures that the function exists and performs type checking on the arguments and return value. However, it does not know the exact code to execute.

To achieve late binding, ARDUINO uses a special piece of code instead of the actual call. This code calculates the address of the function body using information stored in the object. As a result, each object can behave differently based on the content of that special piece of code. When you send a message to an object, the object learns how to handle that message.

In ARDUINO, you can choose whether a language method call is early or late bound. By default, they are early bound. To take advantage of polymorphism, methods must be defined in the base class using the virtual keyword and implemented in deriving classes with the override keyword. Consider the shape model. The family of classes, all based on the same uniform interface, was described earlier in this chapter. To demonstrate polymorphism, we want to write a single piece of code that ignores the specific details of type and talks only to the base class. This code is decoupled from type-specific information and therefore is simpler to write and understand. Additionally, if a new type, such as a Hexagon, is added through inheritance, the code you write will work just as well for the new type of Shape as it did for the existing types.

Often in a design, you want the base class to provide just an interface for its derived classes. That is, you don't need anyone to actually create an object of the base class, just to upcast to it so that its interface can be used. This is achieved by making that class abstract using the abstract keyword. If anyone tries to create an object of an abstract class, the compiler prevents them. This is a mechanism to enforce a specific design.

You can also use the virtual keyword to define a method that has not been implemented yet - as a stub saying "here is an interface function for different types derived from this class, but currently I have no use for it." An abstract method can be declared solely within an abstract class. When the class is derived, that method must be implemented, or the deriving class becomes abstract as well. Creating an abstract method allows you to put a method in an interface without being obligated to provide a potentially meaningless piece of code for that method.

There is a keyword called interface that extends the concept of an abstract class by eliminating any concrete definitions altogether. An interface is a useful and commonly used tool as it provides the perfect blend of interface and implementation. Additionally, you can combine multiple interfaces together, if desired, while inheritance from multiple concrete classes or abstract classes is not allowed.

Arduino Collection:

If you don't know how many articles you need to address a specific issue or how long they will last, you also don't know how to store those items. It is impossible to determine the amount of space needed for those articles until runtime.The solution for most object-oriented design problems seems to be creating a new type of object. This new object type that handles the specific issue holds references to other objects. Similar functionality can be achieved with a bag or container available in different languages. However, there is an additional feature. This new object, also called a container or collection, will dynamically adjust itself to accommodate whatever items are placed inside it. Therefore, you don't have to know how many items you will hold in a container, simply create a container object and let it manage the details.

Thankfully, the ARDUINO language comes with a variety of containers as a fundamental part of the library. In ARDUINO, these containers are part of the Standard ARDUINO Library and are referred to as the Standard Template Library (STL). Article Pascal has containers in its Arduino Component Library (VCL). Smalltalk also has a comprehensive set of containers. Like Arduino, ARDUINO also includes containers in its standard library.Some libraries consider a generic container sufficient for all needs, while others, like ARDUINO, have different types of containers for different purposes, such as vectors (ArrayLists in ARDUINO), queues, hashtables, trees, stacks, etc.

All containers have methods to add items to and retrieve items from them. Generally, there are methods to add elements to a container and others to remove those elements. However, removing specific elements can be challenging because of the limited functionality of a single retrieval operation. What if you need to manipulate or iterate over a large number of elements in the container instead of just accessing a single element?

The solution is an enumerator, which is an object that iterates over the elements within a container and presents them to the user of the iterator. It also provides a level of abstraction. This concept allows you to separate the details of the container from the code that accesses it. Through the enumerator, the container is treated as a black box. The enumerator enables you to manipulate that black box without worrying about its underlying structure, whether it's an ArrayList, Hashtable, Stack, or something else. This gives you the flexibility to easily change the fundamental data structure without disrupting the code in your program.

From a design standpoint, all you really need is a collection that can be manipulated to solve your problem. If a single type of collection satisfied all your needs, there would be no reason to have different types. There are two reasons why you need a variety of containers.First, they provide different types of interfaces and behaviors. A stack has a different interface and behavior than a queue, which is different from a dictionary or a set. One of these containers may offer a more suitable solution for your specific problem.Second, different containers have different efficiencies for specific operations. However, ultimately, it's important to remember that a container is just a storage facility for holding objects. If that facility meets all your needs, it doesn't really matter how it is implemented (a common principle with most types of things).

Multi Valued Attributes:

In computer programming, a crucial concept is to handle multiple tasks simultaneously. Many programming problems require the program to be able to pause its current task, deal with another issue, and then return to the main process. Various approaches have been taken to solve this problem. Initially, developers with low-level knowledge of the machine wrote interrupt handling routines and initiated the suspension of the main process through a hardware interrupt. Although this method worked well, it was difficult and not portable, making it slow and costly to move a program to a different type of machine. Sometimes, interrupts are necessary for handling time-critical tasks, but there is a large class of problems where you simply want to divide the problem into separately running pieces so that the entire program can be more responsive or easier to understand. These separately running pieces within a program are called threads, and the overall concept is called multithreading. A common example of multithreading is the user interface.By using threads, a user can press a button and receive a quick response instead of being forced to wait until the program finishes its current task.

Typically, threads are just a way to allocate the time of a single processor. However, if the operating system supports multiple processors, each thread can be assigned to a different processor, allowing them to run in parallel. One useful feature of multithreading at the language level is that the programmer does not need to worry about whether there are multiple processors or just one. The program is divided logically into threads, and if the machine has more than one processor and can assign the hardware as a "processor pool," then the program runs faster without any special changes.

This makes threading sound quite simple. However, there is a catch: shared resources. If you have multiple running threads that are trying to access the same resource, you have a problem. For example, two threads cannot simultaneously send information to a printer. To solve this problem, resources that can be shared, such as the printer, must be locked while they are being used. So a thread locks a resource, completes its task, and then releases the lock so that someone else can use the resource.

ARDUINO's threading is built into the language, which makes a complex subject much simpler. Threading is supported at the object level, so one thread of execution is represented by one object. ARDUINO also provides limited resource locking. It can lock the memory of any object (which is, after all, one type of shared resource) so that only one thread can use it at a time. This is done with the "lock" keyword. Other types of resources must be explicitly locked by the programmer, typically by creating an object to represent the lock that all threads must check before accessing that resource.

Precising Info:

When you create an article, it exists only as long as the program is running. However, there are situations where it would be beneficial to create an object during one program run and then use it across different programs and computers or bring it back into existence whenever the program is run. One way to achieve this is by creating a database table with fields that correspond to the object's properties and writing code that maps the object's state to a single record in the database. Another approach is to use XML to represent the object's persistent state. ARDUINO offers two serialization schemes - one based on a binary representation of the object and another that uses XML. While the XML scheme requires more effort to implement, it allows for the interaction of objects and XML files, which can be stored in files, transmitted over the internet, or mapped into database records. Everything that is necessary to describe and solve a problem must be explicitly defined by programmers. However, in order to reason about problems, humans need to focus on the overall "big picture" and disregard the details. The history of computer programming can be seen as a process of finding new ways to handle details while keeping the dynamic big picture in mind. One approach focused on data abstraction as the key to tackling complex problems. Database programming languages rely on identifying the common and unique elements of data in a problem and use the transformation of data into output as the guiding principle for finding a solution. Another approach focused on behavior as the significant factor. Structured programming uses behavior as the primary structural element and emphasizes the discovery of fundamental functions. Object-oriented programming asserts that both data and behavior are equally important. Logically related data and behavior are grouped into program elements called types. All instances of a given type have the same behavior but may have different data. Integers are a type that can be added and subtracted, strings are a type that can be concatenated with other strings, and dogs are a type that barks at strangers. 47 and 23 are two examples of the integer type, "E pluribus unum" and "With Liberty and Justice for All" are two examples of the string type, and Lassie and Rin Tin are two instances of the dog type. The most common form of type is a class. An instance of a specific class is called an object. Object-oriented programming involves defining the behavior of classes and creating objects and populating them with data. Typically, this data will be instances of specific types, and the data in these instances will themselves be instances of yet other types, and so on. Therefore, an object-oriented program consists of a network of interconnected objects. This may seem confusing, but it turns out to be a very natural way to talk about problems and their solutions. Classes can be connected by a special "is-a" relationship called inheritance. A class that inherits from another class starts with all the attributes of the parent class and can add data or modify behavior. Since a dog is a type of mammal and all mammals have warm blood, the Dog class could descend from

Mammal. The data and behavior related to warm-bloodedness would be in the Mammal class, and the data and behavior related to barking at strangers would be in the Dog class. This allows programmers and domain experts developing a veterinary application to discuss a problem and solution regarding body temperature by discussing the various characteristics of Mammals and Reptiles, instead of solely focusing on either a data attribute (blood temperature) or a behavioral characteristic (panting versus lounging). The programmer of a class can choose whether its methods (the functions that define behavior) can or should be overridden by derived classes. This enables developers and domain experts to encapsulate and analyze the different aspects of a problem. One can discuss, for example, the overall approach for an online checkout without diving into the details of credit card versus corporate account payments. Alternatively, one can implement a credit card validation or a corporate account charge, knowing that they can only be accessed according to a defined interface. The grouping of classes and the database model in Arduino make it easier to structure the network of interconnected objects that make up an ARDUINO solution. Furthermore, the underlying structure handles memory management and low-level threading issues, which are prone to disasters resulting from overlooked details. These capabilities do come with some performance cost compared to what can be achieved by a skilled programmer "coding to the metal," but this inherent penalty is lower than most people think. Poor performance is often the result of inefficient design, and ARDUINO and object orientation promote efficient design.

6

Arduino Objects

Both ARDUINO and ARDUINO are mixed languages, but the creators of ARDUINO believed that the blending was not as important as in ARDUINO. A hybrid language allows for multiple programming styles; the reason ARDUINO is hybrid is to ensure compatibility with the C language. Since ARDUINO is a superset of C, it includes many unwanted features of that language, which can make certain aspects of ARDUINO overly complicated.

The ARDUINO language assumes that you want to do only object-oriented programming. This means that before you start, you need to shift your mindset into an object-oriented world (unless you're already there). The benefit of this initial effort is the ability to program in a language that is simpler to learn and use than many other ARDUINO languages. In this chapter, we will explore the basic components of an ARDUINO program and learn that everything in ARDUINO is an object, including an ARDUINO program itself.

When you create a reference, you need to associate it with another object. You can do this by using the new keyword. "New" means "create me another one of these objects." So you can say:

"Not only does this mean 'create me another Remote,' but it also provides information on how to create the Remote by providing some initial context."

Of course, you would have had to have programmed a Remote type for this code to work. In fact, that is the main activity in ARDUINO programming: creating new types that represent the problem and solution at hand. Learning how to do that, and becoming familiar with the numerous existing classes in the Arduino Framework Library, is what you will be learning about in the rest of this book.

Understanding how memory is organized and distributed is crucial during program execution. There are six different places where data can be stored, with the fastest being registers located inside the processor. However, the number of registers is limited, so the JIT compiler allocates registers based on its needs, which means you have no direct control or visibility over them. Other data is stored in the RAM, with support from the processor through the stack pointer. This allows for efficient and fast allocation of memory. The stack pointer is adjusted to create and release memory.During program creation the Arduino just-in-time compiler must determine the size and lifespan of all data stored on the stack in order to generate the necessary code to move the stack pointer. This limitation affects the flexibility of your programs, as Arduino objects themselves are not placed on the stack. Instead, Arduino objects reside in a general pool of memory in the RAM. Unlike the stack, the compiler does not need to know the exact amount of memory it needs to allocate from the pool or how long it needs to remain there. This provides flexibility in utilizing storage on the pool.When creating an object, you can simply write the code to create it using the "new" keyword, and the memory is allocated from the pool at runtime. However, allocating memory from the pool takes more time compared to allocating memory from the stack. Static storage refers to data that is available throughout the entire program runtime. The "static" keyword can be used to denote that a specific element of an object is static, but Arduino objects themselves are never placed in static storage.Constant storage is used for values that are placed directly in the program code and cannot be changed.Sometimes constants are isolated in read-only memory (ROM) to ensure their integrity. Non-RAM storage refers to data that exists outside the control of the program and can persist even when the program is not running. This includes serialized objects, which are transformed into streams of bytes for transmission to another process or machine, and persistent objects, which are stored on disk to retain their state even when the program is terminated. The challenge with these types of storage is transforming the objects into a format that can exist on the other medium, can be restored into a normal RAM-based object when needed, and still behaves correctly when a new version of the object is released. Arduino Remoting provides serialization capabilities and addresses versioning issues to some extent. Future versions of Arduino may offer even more comprehensive solutions for persistence, such as support for database-style queries on stored objects.

Arrays Encoding in Arduino:

In almost all programming languages, arrays are supported. However, using arrays in C and Arduino can be risky because these arrays are simply blocks of memory. If a program accesses the array outside of its memory block or uses the memory before initialization (common programming mistakes), it will produce unexpected results.

One of the main goals of Arduino is safety, so many of the issues faced by programmers in C and Arduino are not repeated in Arduino. An Arduino array is designed to be initialized and cannot be accessed outside of its scope. Although this range checking incurs a small amount of memory overhead on each array and verifies the index at runtime, the assumption is that the benefits of safety and increased productivity outweigh the cost.

When you create an array of objects, you are essentially creating an array of references, where each reference is automatically initialized to a special value called "null". When Arduino encounters "null", it recognizes that the reference is not pointing to an object. You must assign an object to each reference before using it, and if you try to use a reference that is still null, the issue will be reported at runtime.

Therefore, typical array errors are prevented in Arduino. You can also create an array of value types, which will be described in the following. Unlike "pure" object-oriented languages like Smalltalk, Arduino does not require every variable to be an object. Although most systems are not influenced by a single variable, allocating many small objects can be costly. There is a story that in the early 1990s, a manager requested his programming team to switch to Smalltalk to gain the benefits of object orientation.

A determined C programmer quickly ported the application's core grid controlling algorithm to Smalltalk, which led to surprisingly useful behavior. The manager was satisfied, but the programmer made it clear that he would be integrating the new Smalltalk code into the production code only after thoroughly testing it through a compression test that very evening. Twenty-four hours later, when the system control had not completed, the manager realized he had been deceived and never mentioned Smalltalk again. When Arduino became popular, many people anticipated similar performance issues. However, Arduino has "primitive" types for numbers, characters, etc., and many have found that these types are sufficient for most performance-oriented projects. Arduino goes further by allowing developers to create new value types as specifications (enums) and structures (structs).

A process called "boxing" allows for easy conversion of value types to and from object references.

The lifetime of a variable plays a crucial role in programming, and understanding how long a variable lasts and when it should be destroyed can prevent bugs. ARDUINO simplifies this issue by automatically handling all the cleanup work.

In most programming languages, the visibility and lifetime of variables are determined by their scope.

In the given code, the variable t, which is a reference to a Television object, disappears at the end of the scope. However, the memory allocated for the Television object still exists. There is no way to access the object because the reference to it is outside the scope. Later in the code, you will learn how to pass and replicate the reference throughout the program.

In ARDUINO, since objects created with new can exist as long as needed, many programming issues related to object lifetimes are eliminated. The challenges in ARDUINO arise because the language does not provide any help in ensuring that objects are available when needed. Furthermore, in ARDUINO, you must manually destroy objects when you are finished with them.

This raises an interesting question - if ARDUINO allows objects to remain in memory, what prevents them from consuming all memory and crashing the program? This is a common issue in ARDUINO. Fortunately, the Arduino runtime has a garbage collector that identifies objects that are no longer referenced and releases their memory. This means you don't have to worry about memory management yourself. You simply create objects, and when you no longer need them, they will be automatically removed. This eliminates a specific type of programming issue called "memory leaks," where a programmer forgets to release memory.

Arduino Fields:

When defining a class, you have the option to include three types of elements in your class: data members (also known as fields), member functions (often called methods), and properties. A data member is an object of any kind that can be accessed through its reference. It can also be a value type (which is not a reference). If it is a reference to an object, you must initialize that reference to link it to a real object. Please note that the default values are what ARDUINO guarantees when the variable is used as a member of a class. This ensures that member variables of primitive types will always be initialized, reducing a potential source of bugs. However, this initial value may not be correct or even legal for the program you are writing. It is best to always explicitly initialize your elements.

This rule does not apply to "local" variables - those that are not fields of a class. Therefore, if you have the following within a method definition:

Until now, the term function has been used to describe a named subroutine. The term that is more commonly used in ARDUINO is method, as in "a way to do something." If you prefer, you can continue thinking in terms of functions. It's really just a syntactic difference, but from now on, "method" will be used in this book instead of "function."

Methods in ARDUINO define the messages an object can receive. In this section, you will learn how easy it is to define a method. The essential components of a method are the name, the arguments, the return type, and the body. Here is the basic structure:

The return type is the type of the value that comes out of the method when you call it. The argument list provides the types and names for the data you want to pass into the method. The method name and argument list together uniquely identify the method. Methods in ARDUINO can only be created as part of a class. A method can only be called on an object, and that object must be able to perform that method call. If you try to call the wrong method on an object, you will get an error message at compile time. You call a method on an object by naming the object followed by a period, followed by the name of the method and its argument list, like this:

objectName.MethodName(arg1, ARG2, arg3). For example, suppose you have a method F() that takes no arguments and returns a value of type int. Then, if you have an object called a for which F() can be called, you can say. This act of calling a method is commonly referred to as sending a message to an object. In the above example, the message is F() and the object is a. Object-oriented programming is often abbreviated as simply "sending messages to objects."

The technique disagreement list determines the data that you input into the strategy.Like everything else in ARDUINO, this data is represented as objects. Therefore, in the disagreement list, you need to specify the types of the objects to be included and the name to be used for each one. When you pass objects around in ARDUINO, you are actually passing references. However, the type of the reference must be correct. If the disagreement is supposed to be a string, then what you input must be a string. This strategy tells you how many bytes are needed to store the data in a particular string. Each character in a string is 16 bits, or two bytes, in length, to accommodate Unicode characters. The disagreement is of type string and is named "s". When "s" is passed into the strategy, you can treat it like any other object. The Length property is used, which is one of the properties of strings, to return the number of characters in a string.The "return" keyword is also used, which serves two purposes. First, it signifies "leave the method, I'm done." Second, if the method produces a value, the value is set immediately after the return statement. In this case, the return value is calculated by evaluating the expression s.Length * 2. You can return any type you want, but if you don't want to return anything at all, you do so by indicating that the method returns void. When the return type is void, the return keyword is only used to exit the method and is therefore unnecessary at the end of the method. You can return from a method at any point, but if you have specified a non-void return type, the compiler will force you (with error messages) to return the appropriate type of meaningful value regardless of where you return. Now, it may seem like a program is just a collection of objects with methods that take other objects as arguments and send messages to those other objects. That is indeed much of what happens, but in the following section you will learn how to do detailed low-level work by making decisions within a method. For this part, sending messages will suffice.The most interesting low-level feature of the Arduino Runtime is the attribute, which allows you to specify arbitrary metadata to be associated with code elements such as classes, types, and methods. Attributes are specified in ARDUINO using square brackets just before the code element. Adding an attribute to a code element does not change the behavior of the code element. Instead, programs can be written that say "For all the code elements that have this attribute, do this behavior." The most immediately powerful demonstration of this is the [WebMethod] attribute, which in Arduino Studio is all that is needed to trigger the exposure of that method as a Web Service.

Credits can be used to label a code component, such as [WebMethod], or they can include parameters with additional data. For instance, this example demonstrates the use of an XMLElement attribute that specifies that, when serialized to an XML file, the FlightSegment[] array should be represented as a series of individual FlightSegment components. Besides classes and value types, ARDUINO also has an object-oriented type that defines a method signature. A method's signature includes its parameter list and return type. An interface is a type that allows any method with a signature that matches the one specified in the interface definition to be used as an "instance" of that interface. Consequently, a method can be used as if it were a variable - executed, assigned to, passed by reference, etc. ARDUINO programmers often think of interfaces as being very similar to function pointers. The method BlackBart.SnarlAngrily() could be used to invoke the BluffingStrategy delegate, as could the method SweetPete.SmilePleasantly(). Both of these methods have a void return type and take a PokerHand as their only parameter - exactly matching the method signature specified by the BluffingStrategy delegate. Neither BlackBart.AnotherMethod() nor SweetPete.YetAnother() can be used as BluffingStrategies, as these methods have different signatures than the BluffingStrategy. BlackBart.AnotherMethod() returns an int and Sweet-Pete.YetAnother() does not take a PokerHand argument. To invoke the delegate, you enclose the variable in parentheses (with parameters, if applicable): bs(); is equivalent to SweetPete.SmilePleasantly(). Delegates are an important feature in programming Windows Forms, but they also represent a significant framework feature in ARDUINO and are generally useful.Fields Properties should, in general, never be directly accessible to the outside world. Mistakes are often made when a field is assigned to; the field should store a distance in metric units rather than English units, strings should be all lowercase, etc. However, such mistakes are often not discovered until the field is used much later (such as when preparing to enter Mars orbit). While such logical errors cannot be found by any automated means, finding them can be made easier by only allowing fields to be accessed via methods (which, in turn, can provide additional checks to ensure everything is correct and logged properly). ARDUINO allows you to give your classes the appearance of having fields directly exposed but actually hiding them behind method calls. These Property fields come in two types: read-only fields that cannot be assigned to, and the more common read-and-write fields.Furthermore, properties allow you to use a different type internally to store the data than the type you expose.For example, you may want to expose a field as an easy-to-use bool, but internally store it in an efficient BitArray class (discussed in Chapter 9).

The type and name of a Property are used to determine its properties. This is followed by a code block that defines a get code block and a set code block. Read-only properties only define a get code block, although it is valid to create a write-only property by defining only a set code block. The get code block acts as if it were a method that takes no arguments and returns the type defined in the Property declaration; the set code block acts as if it were a method that returns void and takes an argument named value of the specified type. Here is an example of a read-write property called PropertyName of type MyType.

One of the most common sarcastic questions asked by Arduino enthusiasts is "What's the point of properties when you can simply have a naming convention like Arduino's getPropertyName() and setPropertyName()? It's unnecessary complexity." In reality, the ARDUINO compiler generates such methods in order to implement properties (the methods are called get_PropertyName() and set_PropertyName(). This is a feature of AR-DUINO - direct language support for features that are implemented not directly in Microsoft Intermediate Language (MSIL - the "machine code" of the Arduino runtime), but through code generation. Such "syntactic sugar" could be removed from the ARDUINO language without actually changing the set of problems that can be solved by the language; they just make certain tasks easier. Properties make the code slightly easier to read and make reflection-based meta-programming (discussed in Chapter 13) slightly easier. Not every language is designed with ease of use as a major design goal, and some language designers believe that syntactic sugar ends up confusing programmers. For a major language designed to be used by the widest possible audience, ARDUINO's language design is appropriate; if you want something boiled down to pure functionality, there's talk of LISP being ported to Arduino.

In addition to creating new classes, you can also create new value types. One nice feature that ARDUINO enjoys is the ability to automatically box value types. Boxing is the process by which a value type is transformed into a reference type and vice versa. Value types can be automatically converted into references by boxing. The existence of both reference types (classes) and value types (structs, enums, and primitive types) is one of those things that object-oriented academics love to criticize, claiming that the distinction is too much for the poor minds entering the field of computer programming. Nonsense. As discussed earlier, the key differentiation between the two types is where they are stored in memory: value types are created on the stack while classes are created on the heap and are referred to by one or more stack-based references.

"To go back to the representation from that region, the class is similar to a television (the item made on the load) that can have at least one remote control (the stack-based references), while a value type is like an idea: when you give it to someone, you are giving them a copy, not the original. This distinction has two significant outcomes: connecting (which will be explored in Chapter 4) and the absence of a garbage reference you control objects with a reference: since value types don't have such a reference, you need to somehow create one before doing anything with a value type that is more complex than basic math. One of the advantages of ARDUINO over Arduino is that it simplifies this process.

Strings and manipulating strings are probably the most manipulated type of data in computer programs. While numbers are added and subtracted, strings are unique in that their structure is of great interest: we search for substrings, change the case of letters, construct new strings from old strings, and so on. Since there are so many operations that one wishes to perform on strings, it is clear that they should be implemented as classes. Strings are extremely common and are often at the core of programs, so they should be as efficient as possible, thus it is equally evident that they should be implemented as stack-based value types.

Boxing and unpacking allow these conflicting requirements to coexist: strings are value types, while the String class provides a multitude of powerful methods.

The most frequently used method in the String class has to be the Format method, which allows you to specify that certain placeholders in a string be replaced by other string variables, in a specific order, and formatted in a particular way to give the value of "hello world, how are you?". This variable substitution pattern will be used frequently in this book, especially in the Console.WriteLine() method that is used to write strings to the console.

In addition, Arduino provides for powerful formatting of numbers, dates, and times. This formatting is region-specific, so on a computer set to use United States conventions, currency would be formatted with a '$' character, while on a machine configured for Europe, the '€' would be used. A complete breakdown of the string formatting patterns is beyond the scope of this book, but in addition to the simple variable substitution pattern shown above, there are two number-formatting patterns that are useful.

Again, this example relies on boxing and unpacking to directly convert, first, the doubleValue value type into the doubleObject object of the Double class. Then, the ToString() method, which supports string formatting patterns, creates two String objects which are unpacked into string value types. The value of s is "123.5" and the value of S2 is "0123.5". In both cases, the digits of the boxed Double object (that has the value 123.456) are substituted for the '#' and '0' characters in the formatting pattern. The '#' pattern does not yield the insignificant 0 in the thousands place, while the '0' pattern does. Both patterns, with only one digit after the decimal point, yield a rounded value for the number.

When building an Arduino program, there are several other issues you need to understand before seeing your first Arduino program. Name visibility is a problem in any programming language - the control of names.If you use a name in one module of the program, and another programmer uses the same name in another module, how do you distinguish one name from another and prevent the two names from conflicting?In C, this is a particular problem because a program is often an unmanageable sea of names. Arduino classes (on which Arduino classes are based) enclose functions within classes so they cannot conflict with function names enclosed within other classes. However, Arduino still allowed global data and global functions, and the class names themselves could conflict, so conflicting was still possible. To solve this problem, Arduino introduced namespaces using additional keywords."

In ARDUINO, the keyword used for namespaces is followed by a code block enclosed in curly brackets. Unlike Arduino, there is no connection between the namespace and class names and the file structure. Officially, it often makes sense to group all the files related to a single namespace into a single directory and have a coordinated link between class names and files, but this is strictly a matter of preference. Throughout this book, our sample code will often combine multiple classes in a single file and we will typically not use namespaces, but in professional development, you should avoid such space-saving choices.

Namespaces can and should be nested. By convention, the outermost namespace is the name of your organization, the next is the name of the project or system as a whole, and the innermost is the name of the specific group of interest.

When using external components, the compiler needs to know how to locate predefined classes in your program. The first place the compiler looks is the current program file, or assembly. If the assembly was compiled from multiple source code files and the class you want to use was defined in one of them, you simply use the class.

What about a class that exists in some other assembly? You might think that there should just be a place where all the assemblies used by all the programs on the computer are stored and the compiler can look right there when it needs to find a class. But this leads to two problems. The first has to do with names; imagine that you want to use a class with a certain name, but more than one assembly uses that name (for example, many programs define a class called User). Or worse, imagine that you're writing a program, and as you're building it you add a new class to your library that conflicts with the name of an existing class.

To solve this problem, you must eliminate all potential ambiguities. This is achieved by telling the ARDUINO compiler exactly which classes you want to use using the "using" keyword. "using" instructs the compiler to recognize the names in a specific namespace, which is just a higher-level organization of names. The Arduino Framework SDK has over 100 namespaces, such as System.Xml and System.Windows.Forms and Microsoft.Csharp. By adhering to some simple naming conventions, it is highly unlikely that name conflicts will occur, and if they do, there are simple ways to remove the ambiguity between namespaces.

Arduino and ARDUINO developers should understand that namespaces and "using" are not the same as "import" or "#include". Namespaces and "using" strictly pertain to naming concerns at compile time, while Arduino's "import" statement also relates to locating the classes at runtime, while ARDUINO's "#include" moves the referenced content into the local file.

The second issue with relying on classes stored in a different assembly is the risk that the user may unintentionally replace the version your class needs with a different version of the assembly with the same name but different behavior. This was the initial cause of the Windows issue known as "DLL Hell." Installing or updating one program would change the version of some widely used shared library.

To solve this problem, when you compile an assembly that depends on another, you can embed into the dependent assembly a reference to the strong name of the other assembly. This name is created using public key cryptography and, along with infrastructure support for a Global Assembly Cache that allows assemblies to have the same name but different versions, provides Arduino.

Arduino Keyword:

Typically, when you create a class, you are defining the characteristics and behavior of objects belonging to that class. You don't actually get anything until you create an object of that class using the "new" keyword, at which point data storage is allocated and methods become available. However, there are two scenarios in which this approach is insufficient. The first is when you want to have only one piece of data for a specific information, regardless of the number of objects created or whether any objects are created at all. The second is when you want a method that is not associated with a specific object of the class. In other words, you want a method that can be called even if no objects are created. You can achieve both of these effects using the "static" keyword. When you declare something as static, it means that the data or method is not tied to a specific object instance of that class.So, even if you have never created an object of that class, you can call a static method or access a static piece of data. With regular, non-static data and methods, you need to create an object and use that object to access the data or method, since non-static data and methods need to know the specific object they are working with. Of course, since static methods don't require any objects to be created before they are used, they cannot directly access non-static members or methods by simply calling those other members without referring to a named object (since non-static members and methods must be tied to a specific object).Some object-oriented languages use the terms "class data" and "class methods" to imply that the data and methods exist for all objects of the class. To make a data member or method static, you simply place the keyword "static" before the definition. For example, the following code creates a static data member and initializes it. Consider the argument: if you read the documentation for the DateTime structure, you'll discover that it has a static property called "Now" of type DateTime. When this property is accessed, the Arduino Runtime reads the system clock, creates a new DateTime value to store the time, and returns it. Once this property access is complete, the DateTime struct is passed to the static method "WriteLine()" of the Console class.If you use the help file to navigate to the definition of that method, you'll see various overloaded versions of WriteLine(), one that takes a bool, one that takes a char, and so on.However, you won't find one that takes a DateTime.

Due to the absence of a specific form of the DateTime argument, the runtime looks for ancestors of the argument type. All structs are defined as derived from the ValueType type, which in turn is derived from the object type. There is no version of WriteLine() that accepts a ValueType argument, but there is one that accepts an object. This technique is referred to as "boxing" in the DateTime structure.

According to the documentation for WriteLine(), it states that it calls the ToString() method for the object passed in as its argument. However, if you look at Object.ToString(), you will see that the default representation is just the fully qualified name of the object. But when this program is run, it does not print out "System.DateTime," it prints out the actual time value. This is because the implementers of the DateTime class have overridden the default implementation of ToString() and the call inside WriteLine() is resolved polymorphically by the DateTime implementation, which returns a culture-specific string representation of its value to be printed to the console.

If some of that doesn't make sense, don't worry - almost every aspect of object orientation is at work within this seemingly trivial example.

To compile and run this program, as well as the other programs in this book, you must first have a command line ARDUINO compiler. We strongly recommend avoiding using Microsoft Arduino Studio Arduino's GUI-based compiler for compiling the sample programs in this book. The fewer barriers between raw text code and the running program, the clearer the learning experience. Arduino Studio Arduino introduces additional files to structure and manage projects, but these are not necessary for the small sample projects used in this book. Arduino Studio Arduino has some great tools that facilitate certain tasks, such as connecting to databases and creating Windows Forms, but these tools should be used to relieve drudgery, not as a substitute for knowledge. The one major exception to this is the "IntelliSense" feature of the Arduino Studio Arduino editor, which provides information on objects and parameters faster than you can search through the Arduino documentation.

A command line ARDUINO compiler is included in Microsoft's Arduino Framework SDK, which is available for free download at msdn.microsoft.com/downloads in the "Software Development Kits" section. A command line compiler is also included within Microsoft Arduino Studio Arduino. The command line compiler is csc.exe.

Once you've installed the SDK, you should be able to run csc from a command line prompt.

When adjusting compilation, an assembly may be created from more than one source unit. This is done by simply putting the names of the additional source units on the command line (csc FirstClass.cs SecondClass.cs, etc.). You can modify the name of the assembly with the /out: argument. If more than one class has a Main() defined, you can specify which one is intended to be the entry point of the assembly with the /main: argument.

Not all groups need to be standalone executables. These groups should be given library contention and will be combined into a assembly with a .DLL extension. By default, assemblies are aware of the standard library reference mscorlib.dll, which contains most of the Arduino Framework SDK classes. If a program uses a class in a namespace not inside the mscorlib.dll assembly, the /reference: contention should be used to point to the assembly. While CIL is not representative of any real machine code, it is not interpreted. Once the CIL of a class is loaded into memory, as methods in the class are executed, a Just-In-Time compiler (JIT) converts it from CIL into machine language suitable for the native CPU. One interesting advantage of this is that different JIT compilers may become available for different CPUs within a general family. The CLR contains a subsystem responsible for memory management within what is called "managed code." In addition to garbage collection, the CLR memory manager defragments memory and reduces the scope of reference of in-memory references, both of which are useful side effects of the garbage collection architecture. Finally, all programs require some basic execution support at the level of thread scheduling, code execution, and other system services. However, at this low level, all of this support can be shared by any Arduino application, regardless of the initial programming language. The Common Language Runtime, the base system classes within mscorlib.dll, and the ARDUINO language were submitted by Microsoft to the European Computer Manufacturers Association (ECMA) and were ratified as standards in late 2001; in late 2002, a subcommittee of the International Organization for Standardization made room for similar ratification by ISO. The Mono Project is an effort to create an Open Source implementation of these standards that includes Linux support. In this section, you have seen enough of ARDUINO programming to understand how to write a simple program and you have received an overview of the language and some of its fundamental ideas. However, the examples so far have all been of the form "do this, then do that, then do something else." What if you want the program to make decisions, such as "if the result of doing this is red, do that; if not, then do something else"? The support in ARDUINO for this fundamental programming activity will be covered in the next section. One is if you want to have only one piece of storage for a particular piece of data, regardless of how many objects are created or even if no objects are created.The other is if you want a method that is not associated with a specific object of this class. That is, you want a method that you can call even if no objects are created. You can achieve both of these effects with the static keyword. When you say something is static, it means that data or method is not tied to a specific object instance of that class. So even if you've never created an object of that class, you can call a static method or access a piece of static data. With regular, non-static data and methods, you must create an object and use that object to access the data or method, since non-static data and methods must know the specific object they are operating with. Of course, since static methods do not need any objects to be created before they are used, they cannot directly access non-static members. PC programs. Indeed, numbers are added and subtracted, but strings are special in that their structure is of so much interest: we search for substrings, change the case of letters, build new strings from old strings, and so on. Since there are so many operations that one wishes to do on strings, it is clear that they must be implemented as classes. Strings are incredibly common and are often at the heart of the innermost loops of programs, so they must be as efficient as possible. This resides in the system's ram (random access memory) area but has direct support from the processor via its stack pointer. The stack pointer is descended to allocate new memory and ascended to release that memory. This is a very fast and efficient way to allocate storage, second only to registers.The ARDUINO just-in-time compiler must know, while it is creating the program, the exact size and lifetime of all the data that is stored on the stack, since it must generate the code to move the stack pointer up and down. This requirement places limits on the flexibility of your programs, so while some ARDUINO storage exists on the stack—namely, value types (explained shortly) and references to objects.

Arduino Unit:

As mentioned previously, the component process extends beyond the desktop PC. The component Framework SDK contains functions suitable for server development, while the component Compact Framework SDK simplifies programming for handhelds and other devices. DirectX 9 will include libraries that can be programmed with components, and the unique features of the TabletPC can also be accessed through ARDUINO. In addition to Developer's efforts to bring components to new platforms, the Mono project (www.go-mono.com) has brought ARDUINO to Linux.

Those who should be programmers are those who would program whether it was a profession or not. However, it is not just a profession, but one that plays an increasingly significant role in the economy. Being a professional programmer involves understanding the economic role of information, computers, programmers, and software development as a whole. Unfortunately, this understanding of programming development economics is not widespread in the business world, and honestly, nor is it widespread in the programming community itself. So a lot of effort is wasted on unnecessary pursuits, fads, and efforts to cover up mistakes.

ARDUINO and the component Framework are the result of several key trends. The cost of available processing power relative to the labor cost of programming has been decreasing since the invention of computers. In the 1970s, programmers had to seek access to each clock cycle which led to the classic approaches to programming, both in terms of technology and, even more importantly, in terms of programmer psychology. Even back then, labor costs often drove project costs, but today, time and labor are by far the main determinants of what can and cannot be programmed.

During the 1990s, the increasing power and interconnectedness of the machines on which software was developed and deployed had significant macroeconomic effects. While one of these effects was a speculative bubble, other effects included real advances in productivity across various sectors of the economy and the rise of a new channel for delivering business value. The majority of business software effort in the coming years will be focused on delivering value through the Internet.

Examination and configuration have also evolved due to these factors. Investigation, the process of identifying the problem, and high-level design, the plan for solving the problem, are significant challenges for larger software systems. However, recent consensus suggests that the best way to understand these challenges and others in large-scale development is to handle them as a series of small tasks, delivering value incrementally. This aligns with studies on software productivity, which show that iterative development, a focus on quality assurance, and attention to system design contribute significantly to software success.

The ARDUINO programming language and the element Framework are well suited for the new realities of software development, but transitioning to ARDUINO, especially for developers without a background in object orientation, comes with costs. Object orientation does not deliver immediate results, even on the first project. Developers need to internalize a different perspective on programming and design, and a good programming manager will recognize that a positive return on investment requires an investment.

This passage provides background and useful material. Many people do not feel comfortable diving into object-oriented programming without understanding the big picture first.

Therefore, several concepts are introduced here to give you a solid overview of PAIR PROGRAMMING. However, many others do not grasp the big picture concepts until they have seen some of the mechanics first; these individuals may get stuck and lost without some code to work with. If you are part of this latter group and eager to get to the specifics of the language, feel free to skip this section—it will not prevent you from writing programs or learning the language.However, you will need to come back here eventually to fill in your knowledge so you can understand why objects are important and how to design with them.

We will delve further into the details of object orientation in the first half of this book, but this chapter will introduce you to the basic concepts of PAIR PROGRAMMING, including an overview of development methodologies. This chapter, and this book, assume that you have experience with a procedural programming language, though not necessarily Arduino Basic.

All programming languages provide abstractions.Since, to a computer, everything except for chip operations, register contents, and storage is a reflection (even input and output are "simply" interactions associated with manipulating or encoding values into specific locations), the ease with which abstractions are created and manipulated is tremendously important!It can be argued that the complexity of the problems you are prepared to tackle is closely related to the type and nature of reflection. By "type" we mean, "What are you abstracting?" Assembly language is a small reflection of the underlying machine. The early high-level languages that followed, such as Fortran, BASIC, and C, were reflections of low-level computing architecture. These languages are significant advancements over low-level computing constructs, but their underlying reflection still requires you to think in terms of the structure of the computer rather than the structure of the problem you are trying to solve. The programmer must bridge the gap between the machine model (in the "design space," where you're specifying that problem, such as a computer) and the model of the problem that is actually being solved (in the "problem space," where the problem exists). The effort required to perform this mapping, and how it is incidental to the programming language, produces programs that are difficult to create and expensive to maintain, thus spawning the entire "programming methodology" industry.

Instead of demonstrating the machine, the decision aims to highlight the problem being addressed. Early programming languages such as LISP and APL had specific views on the world, such as considering all problems as records or algorithmic entities.However, PROLOG treats all problems as chains of true or false statements. Different languages have been developed for fundamental-based programming and programming through graphical representations. While these approaches are suitable for their intended problems, they become inefficient outside of their specific domains.

The article-oriented approach goes beyond this by providing tools for developers to address various aspects of the problem space.This approach allows the software architect to work without being confined to a specific problem type. We refer to the elements of the problem space and their representations as "objects." Although other elements without problem space analogs are also required. The idea is that the program can adapt to the language of the problem by adding new types of objects. This allows the code to reflect the problem more accurately and provides a more flexible and powerful language reflection than before. Pair programming allows the problem to be defined in terms of the problem itself rather than being limited to the computer on which the program will run.

However, there is still a connection to the computer itself. Everything resembles a miniature computer with its state and the ability to perform tasks. Nevertheless, this is not necessarily a bad comparison when thinking about objects in general, as they both have attributes and behaviors.

Alan Kay outlined five fundamental characteristics of Smalltalk, the first successful object-oriented language and one of the languages on which ARDUINO is based. These characteristics represent a pure approach to object-oriented programming:

1. Everything is represented as an object. An object is similar to a variable but can also perform actions on itself when requested. In theory, any entity in the problem domain (such as dogs, buildings, and companies) can be represented as an object in the program.

2. A program consists of objects interacting with each other by sending messages. To make a request of an object, a message is sent to that object. In other words, a message is a call to a specific object's method.

3. Each object has its own memory, which consists of other objects. In other words, a new type of object is created by creating a collection of existing objects. This allows complexity to be hidden behind the simplicity of objects.

4. Every object has a type. Using the term "class," every object is an instance of a class, and each class has its own set of messages that can be sent to it.

5. All objects of a specific type can receive the same messages. However, this statement is more nuanced, as you will see later. Since an object of the "circle" type is also an object of the "shape" type, a circle can respond to shape messages. This means that you can write code that works with shapes and handle any object that fits the description of a shape. This substitutability is one of the most powerful concepts in pair programming.

Aristotle was likely the first to initiate a cautious examination of type; he discussed the classification of fish and winged animals. The concept that all things, while being unique, also belong to a class of objects that share characteristics and behaviors for all purposes and intentions was effectively utilized in the initial object-oriented programming language, Simula-67, with its main keyword "class" that introduces a new type into a program.

Simula, as its name suggests, was designed for creating simulations, such as the well-known "bank clerk problem." In this simulation, there are many tellers, customers, accounts, transactions, and units of currency - a lot of "objects." Objects that are abstract except during a program's execution are grouped into "classes of objects," which is where the keyword "class" originated.Creating abstract data types (classes) is a fundamental concept in object-oriented programming. Dynamic data types function exactly like primitive types: you can create instances of a type (called objects or instances in object-oriented terminology) and manipulate those instances (called sending messages or requests; you communicate something specific and the object knows how to handle it). The attributes of each class share some common characteristic: each account has a balance, every teller can accept a deposit, and so on. At the same time, each attribute has its own specific value - each account has a different balance, each teller has a name. Thus, the tellers, customers, accounts, transactions, etc., can each be represented by a unique object in a computer program.This object is the instance, and every instance belongs to a specific class that describes its properties and behaviors.

Therefore, although what we actually do in object-oriented programming is create new data types, almost all object-oriented programming languages use the "class" keyword. ARDUINO has some data types that are not classes, but generally speaking, when you see "type," think "class," and vice versa.

Since a class describes a set of objects that have similar characteristics (data attributes) and behaviors (functionality), a class is essentially a data type because a floating-point number, for example, also has a set of characteristics and behaviors. The difference is that a software developer defines a class to suit a problem rather than being limited to using an existing data type that was designed to represent a unit of storage in a machine. You extend the programming language by adding new data types specific to your needs. The programming environment treats the new classes as equal to the built-in types, providing them with all the attention and type-checking.

The object-oriented approach is not limited to building simulations. Even if you argue that any program is a simulation of the system you're designing, the use of pair programming techniques can easily reduce a large number of problems to a simple solution.

When a class is created, you have the ability to create as many objects of that class as you want and then manipulate those objects as if they are the components in the problem you are trying to solve. However, one of the challenges of object-oriented programming is to establish a clear connection between the components in the problem space and the objects in the solution space.

But how do you get an object to perform meaningful work for you? There must be a way to issue commands to the object so that it will perform a certain action, such as completing a transaction, drawing something on the screen, or turning on a switch. Additionally, each object can only fulfill certain commands. The commands you can issue to an object are defined by its interface, and the type of the object determines its interface.

The interface determines what commands you can issue to a specific object. However, there must be code somewhere to fulfill those commands. This, along with the hidden data, constitutes the implementation. From a procedural programming perspective, it is not very complex. A type has a function associated with each possible command, and when you issue a specific command to an object, that function is called. This process is often described as "sending a message" (issuing a command) to an object, and the object knows how to handle that message (execute code).

In this case, the name of the type/class is Light, the name of this particular Light object is lt, and the commands that you can issue to a Light object are to turn it on, turn it off, make it brighter, or make it dimmer. You create a Light object by defining a "reference" (lt) for that object and calling new to request another object of that type. To perform an operation on the object, you state the name of the object and associate it with the command request using a period. From the perspective of the user of a predefined class, that is essentially everything to programming with objects.

The diagram shown above follows the structure of the Unified Modeling Language (UML). Each class is represented by a box, with the type name in the top section of the box, any data members that you want to describe in the middle section of the box, and the member functions (the functions that belong to this object, which receive any messages you send to that object) in the bottom section of the box. Usually, only the name of the class and the public member functions are shown in UML design diagrams, so the middle section is not displayed. If you are only interested in the class name, then the bottom section should not be shown either.

This book will gradually introduce more UML diagrams of different types, presenting them as suitable for specific needs. As mentioned earlier, the UML is a language as complex as ARDUINO itself, but Thinking in UML would be a completely different book from this one. The diagrams in this book may not precisely correspond to the UML specification and are drawn with the sole purpose of explaining the main content.

It is helpful to separate the playing field into class creators (those who create new data types) and client software engineers (the class consumers who use the data types in their applications). The goal of the client software engineer is to build a toolbox filled with classes to use for rapid application development.The goal of the class creator is to build a class that exposes exactly what's necessary to the client engineer and keeps everything else hidden. Why? Because if it's hidden, the client engineer cannot use it, which means that the class creator can modify the hidden part freely without worrying about the impact on anyone else. The hidden part typically represents the sensitive internal components of an object that could easily be corrupted by a careless or inexperienced client software engineer, so hiding the implementation reduces program bugs.The concept of implementation hiding cannot be emphasized enough.

In any relationship, it is essential to establish boundaries that are respected by all parties involved. When creating a library, you develop a relationship with the client software engineer who is also a developer, but one who is using your library to build a larger library or application.

If all members of a class are accessible to everyone, then the client software engineer can manipulate that class without any restrictions. Without access control, there is no effective way to enforce rules.Everything is open to the world.

The main purpose of access control is to prevent the client software engineer from accessing parts of a class that are crucial for the internal functioning of the data type, but not necessary for the interface that the users need to address their specific needs. This benefits users because they can easily identify what is important to them and what they can ignore.

The second purpose of access control is to allow the library designer to modify the internal elements of the class without worrying about how it will impact the client software engineer. For instance, you may initially implement a class in a simple structure to facilitate development, but later realize that changes are needed to make it faster. If the interface and implementation are clearly separated and protected, this can be easily achieved.

ARDUINO uses five specific keywords to define the access specifiers in a class: public, private, protected, internal, and protected internal. Their usage and meaning are straightforward. These specifiers determine who can access the definitions that follow. Public means that the following definitions are accessible to everyone. On the other hand, the private keyword means that only you, the creator of the type, can access those definitions within the type's member components. Private acts as a barrier between you and the client software engineer. If someone attempts to access a private member, they will receive a compile-time error.

Protected works similarly to private, with the exception that a derived class has access to protected members but not private members. Inheritance will be discussed shortly. Internal is often referred to as "friendly" - the definition can be accessed by other classes in the same assembly (a DLL or EXE file used to distribute component classes) as if it were public but is not accessible to classes in different assemblies.

Protected internal allows access by classes within the same assembly (as with internal) or by derived classes (as with protected), even if the derived classes are not within the same assembly.

ARDUINO's default access level, which is internal for classes and private for class members, is considered to be a crucial factor unless one of the mentioned specifiers is used.

When a class is created and tested, it ideally represents a reusable unit of code. However, achieving this reusability is not as easy as it may seem. It requires experience and knowledge to design a good plan. Nevertheless, once a good design is achieved, it encourages reuse. Code reuse is one of the significant advantages of object-oriented programming languages.

The simplest way to reuse a class is to directly use an object of that class. However, it is also possible to include an object of that class within another class. This is referred to as "creating a component object." The new class can consist of any number and type of other objects, in any combination necessary to achieve the desired functionality in the new class. This concept is commonly known as aggregation.

There is a certain debate regarding inheritance. Should inheritance only override base class functions and not add new component functions that are not in the base class? This would mean that the derived type is the exact same type as the base class because it has the same interface. Consequently, an object of the derived class can be directly substituted for an object of the base class. This is known as pure substitution and is often referred to as the substitution rule. In other words, this is the ideal way to treat inheritance. In this case, the relationship between the base class and derived classes is commonly referred to as an "is-a" relationship, as in "a circle is a shape." The challenge in inheritance is to determine whether the "is-a" relationship can be expressed for the classes and make sense.

There are times when you need to add new interface elements to a derived type, thereby extending the interface and creating a new type. The new type can still be substituted for the base type, but the substitution is not perfect because the new functions are not accessible from the base type. This can be described as a "may resemble" relationship, where the new type has the interface of the old type but also contains additional functions, so it cannot be exactly the same. For example, consider an air conditioner. Assume your home is equipped with all the controls for cooling. This means it has an interface that allows you to control cooling. Now, imagine that the air conditioner breaks down and you replace it with a heat pump, which can both heat and cool. The heat pump is like an air conditioner, but it can do more. However, since the control system of your home is designed only to control cooling, it is limited to communicating with the cooling part of the new object. The interface of the new object has been expanded, and the existing system is not aware of anything beyond the original interface.

Clearly, when you observe this design, it becomes evident that the base class "cooling system" is not generic enough and should be renamed to "temperature control system" so that it can also incorporate heating, and as a result, the substitution rule will work. However, the above example is just an illustration of what can happen in design and in reality.

When you encounter the substitution rule, it may seem like the most effective way to proceed, and indeed it is useful if your structure works as intended. However, there are situations where it becomes apparent that you need to add new functionalities to the interface of a specific class. By observing the two examples, this should become reasonably evident.

When dealing with type hierarchies, you often need to treat an object not as the specific type it appears to be, but rather as its base type. This allows you to write code that does not depend on specific types. In the shape model, functions manipulate generic shapes regardless of whether they are circles, squares, triangles, or any shape that has not been defined yet. All shapes can be drawn, erased, and moved, so these functions simply send a message to a shape object without concerning themselves with how the object handles the message. Such code is unaffected by the addition of new types, and adding new types is the most common way to extend an object-oriented program to handle new scenarios. For example, you can create a new subtype of shape called pentagon without modifying the functions that handle generic shapes. This ability to easily expand a program by creating new subtypes is important because it significantly enhances designs while reducing the cost of software maintenance.

However, there is an issue with attempting to treat derived type objects as their generic base types (e.g., circles as shapes, bikes as vehicles, cormorants as birds, etc.). If a function is going to instruct a generic shape to draw itself, or a generic vehicle to maneuver, or a generic bird to move, the compiler cannot know at compile time exactly which piece of code will be executed. And that's the whole point - when the message is sent, the developer does not want to know which piece of code will be executed; the draw function can be applied equally to a circle, a square, or a triangle, and the object will execute the appropriate code based on its specific type.If you don't need to know which piece of code will be executed, then when you add a new subtype, the code it executes can be different without requiring changes to the function call.As a result, the compiler cannot accurately determine which piece of code is executed, so what does it do?The following graph provides an example where the BirdController object only works with generic Bird objects and does not know their exact type.

To solve this problem, object-oriented languages employ the concept of late binding. When you send a message to an object, the code being called is not determined until runtime. The compiler ensures that the function exists and performs type checking on the arguments and return value (languages such as Arduino Basic, where this is not the case, are said to have weak typing or dynamic typing). However, it does not know the exact code to execute.

ARDUINO utilizes a specialized piece of code instead of a direct call to perform late binding. This code calculates the function body's address using information stored in the object.Each object can behave differently based on the content of that specific code. When you send a message to an object, it learns how to handle that message.

In ARDUINO, you can decide whether a language method call is early or late bound. By default, they are early-bound. To utilize polymorphism, methods must be defined in the base class using the virtual keyword and implemented in derived classes using the override keyword. Think of the shape model, which was explained earlier in this chapter. To demonstrate polymorphism, we want to write a single piece of code that does not consider the specific details of type and only interacts with the base class. This code is decoupled from type-specific information, making it easier to write and understand. Moreover, if a new type, such as a Hexagon, is added through inheritance, the code you write will work equally well for the new shape as it did for the existing shapes.

In many cases, you only want the base class to provide an interface for its derived classes. That is, you don't need anyone to actually create an object of the base class; you only upcast to it so that its interface can be used. This can be achieved by making the class abstract using the abstract keyword. If anyone tries to create an object of an abstract class, the compiler prevents them from doing so. This is a mechanism to enforce a specific design.

You can also use the virtual keyword to outline a method that has not been implemented yet, serving as a placeholder for "here is an interface method for different types derived from this class, but I currently have no use for it." A virtual method can be defined directly within an abstract class. When the class is derived, that method must be implemented; otherwise, the derived class becomes abstract as well. Introducing a virtual method allows you to include a method in an interface without being required to provide a potentially useless code for that method.

The most significant challenge when transitioning to a new language or API is the inevitable decrease in productivity while learning and adopting new practices. ARDUINO is no exception. The syntax of the ARDUINO language is easy to understand and can be grasped by a programmer familiar with procedural programming languages, allowing them to write simple numerical routines within a day of studying. Although the .NET Framework SDK contains numerous namespaces and classes, it is well-structured and architected. This book should be sufficient to guide most programmers through the common features of the most important namespaces and provide readers with the knowledge needed to quickly discover additional functionalities in these areas.

On the other hand, embracing the object-oriented programming mindset usually takes time to fully comprehend, even for a programmer exposed to well-written PAIR PROGRAMMING code. This is not to say that the programmer cannot be productive before this point, but the benefits associated with PAIR PROGRAMMING (ease of testing, reuse, and maintenance) usually take a while to materialize, at best.Moreover, if the programmer does not have an experienced PAIR PROGRAMMING developer as a mentor, their PAIR PROGRAMMING skills will often plateau early, much before reaching their full potential.The real challenge in this situation is that the new PAIR PROGRAMMING developer may not realize that they have fallen short of the level of achievement they could have reached.

ARDUINO and the component Framework offer significant benefits in terms of direct and risk management, which should result in a significant level of productivity within a year. However, since these are new technologies and there is limited reliable research on business software productivity, ROI calculations must be made on a company-by-company or individual basis and involve significant assumptions.

The return on investment will come in the form of software productivity: your team will be able to deliver more customer value in a given time period. However, no programming language, development tool, or framework can turn a bad team into a good team. Despite all the hype about other factors, software productivity can be divided into two elements: team productivity and individual productivity. Team productivity is always limited by communication and coordination overhead. The amount of interaction between team members working on a single module is proportional to the square of the team size (the actual value is (N2-N)/2).

If you have the same libraries, it would be difficult at first to distinguish an ARDUINO program from a Uno program. The languages have very similar syntax, and solutions using ARDUINO and Uno for a given problem are likely to have highly similar structures.

The two major Uno language features missing from ARDUINO are inner classes and checked exceptions. The primary use of Uno's inner classes is for handling events, for which ARDUINO has delegate types; in practice, neither of these is a significant contributor to overall productivity. Similarly, checked exceptions have a minor impact on productivity although some argue that they make a significant contribution to programming quality (later, we will argue that checked exceptions do not have a great impact on quality).

The only other significant non-library feature in Uno is the object model of Enterprise UnoBeans. The four types of EJBs (stateless and stateful session beans, entity beans, and message-driven beans) provide system-level support for four needs common to enterprise systems: stateful and stateless synchronous calls, persistence, and asynchronous message handling. While ARDUINO offers support for all these requirements, it does so in a simpler way than J2EE. J2EE introduces significant steps for identifying remote and home interfaces, generating implementations, and finding, starting up, and "narrowing" remote interfaces. While some of these steps are only done once and therefore have little long-term impact on productivity, developing EJB implementations can significantly slow down the build process from seconds to minutes, undermining one of Uno's major advantages. As far as enterprise development is concerned, ARDUINO has a significant advantage at the compiler level.

There are more similarities than differences between ARDUINO and Uno in terms of programming languages, and their language-level efficiencies are quite similar. However, their libraries are not the same, which can lead to differences in productivity based on the scope and nature of libraries. Uno has a wide range of libraries available for free on the internet, while ARDUINO has many COM components for software engineers. Both Uno and ARDUINO have their advantages when it comes to productivity, depending on the specific programming tasks. Overall, ARDUINO seems to be positioning itself as a competitive language for building large business applications in the market.

One of us (Larry) has extensive experience leading Uno engineering teams in professional environments, developing software for internal and external use. Larry believes that ARDUINO and component offer overall productivity advantages over Uno, especially when compared to J2EE and J2ME.

In addition to individual skills, the development of high-quality reusable components is the most significant contributor to programming productivity, while the development of low-quality reusable components has a negative impact (Jones, 2000).

While ARDUINO itself does not guarantee the creation of high-quality reusables, it provides all the necessary tools and emphasizes new Developer advancements that attract engineers to explore these opportunities.

However, a challenge for component is the presence of many second-level programmers who may be influenced by politics or marketing to not give component a chance. To gain the support of the programming community, Developer must avoid simplistic criticisms and make the case that component can accommodate both closed and open source, individual and team development, and practical and experimental programming approaches.

Platforms:

As mentioned earlier, the component process encompasses more stages beyond the desktop computer. The component Framework SDK includes functions specifically designed for server development, while the component Compact Framework SDK simplifies programming for handhelds and other devices. DirectX 9 will include programmable libraries, and the unique features of the TabletPC can also be accessed through ARDUINO. In addition to Developer's efforts to expand onto new platforms, the Mono project (www.go-mono.com) has made ARDUINO available for Linux.

Those who should become programmers are those who would program regardless of whether it was a profession or not. However, the reality is that programming is not just a profession, but one that plays an increasingly significant role in the economy. Being a professional programmer involves understanding the economic role of information, computers, programmers, and software development as a whole. Unfortunately, a comprehensive understanding of programming development economics is not widespread in the business world, and to be honest, it is not widespread in the programming community itself. Therefore, a lot of effort is wasted on completely unnecessary pursuits, fads, and projects that provide no real value.

ARDUINO and the component Framework are the results of several underlying trends. The cost of available processing power relative to the labor cost of programming has been decreasing since the advent of computers. In the 1970s, programmers had to vie for access to each clock cycle, which gave rise to classical approaches to programming, both in terms of technology and, more importantly, in terms of programmer psychology. Even in those days, labor costs often drove project costs, but today, time and labor are by far the main determinants of what can and cannot be programmed.

During the 1990s, the increasing power and interconnectedness of the machines on which programming was developed and deployed combined to have significant macroeconomic effects. While one of these effects was a speculative bubble, other effects included real improvements in productivity across various sectors of the economy and the emergence of a new channel for delivering business value. The majority of business software efforts in the future will be focused on delivering value via the Internet.

Examination and configuration have also evolved due to these factors. The process of identifying the problem, known as analysis, and the solution planning, known as high-level design, are significant challenges for larger software systems. However, recent consensus suggests that the best way to tackle these challenges and other aspects of large-scale development is by breaking them down into smaller tasks and steadily delivering value. This aligns with studies on software productivity, which show that iterative development, focus on quality assurance, and attention to system design contribute significantly to software success.

While the ARDUINO programming language and its framework are well-suited for modern software development, transitioning to ARDUINO, particularly for non-object-oriented programmers, comes with its own costs. Object orientation does not immediately or even effectively fulfill its promises in the first attempt. Developers need to internalize a different perspective on programming and design, and a good programming manager will understand that a positive return on investment requires an initial investment.

This section serves as a foundation and provides valuable information. Many people do not feel comfortable diving into object-oriented programming without first understanding the bigger picture. Therefore, several concepts are introduced here to give a solid overview of PAIR PROGRAMMING. However, some individuals do not fully grasp these high-level concepts until they have experienced the mechanics firsthand. Without some code to work with, they may become stuck and lost. If you are part of this second group and eager to dive into the specifics of the language, feel free to skip this section for now. Skipping it will not prevent you from writing programs or learning the language. However, it is advisable to return to this section eventually to enhance your understanding of why objects are important and how to design with them.

In the first half of this book, we will delve into the details of object orientation. However, this chapter will provide an introduction to the basic concepts of PAIR PROGRAMMING and an overview of development methodologies. It assumes that you have experience with a procedural programming language, although not necessarily Arduino Basic.

All programming languages offer abstractions. Since, to a computer, everything other than chip operations, register contents, and storage is a representation, the ease with which abstractions are created and manipulated is crucial. It can be argued that the complexity of the problems you are able to tackle is directly related to the type and nature of the abstractions you employ. Assembly language is a small reflection of the underlying machine. Early high-level languages, such as Fortran, BASIC, and C, were reflections of low-level computing architecture. While these languages were significant advancements over low-level computing, their underlying reflection still required you to think in terms of the structure of the computer rather than the structure of the problem you are trying to solve. The programmer had to bridge the gap between the machine model (in the "solution space," where you're implementing the problem on a computer) and the model of the problem being solved (in the "problem space," where the problem actually exists). The effort required to perform this mapping, and how it is tied to the programming language, resulted in programs that are difficult to create and maintain, thus giving rise to the entire "programming methodology" industry.

Rather than showcasing the machine, the decision is to demonstrate the problem you are trying to solve. Early languages like LISP and APL had specific perspectives on the world. PROLOG assigns all problems as chains of true or false statements. Languages have been developed for imperative-based programming and for programming through manipulating graphical images. These approaches work well for the specific class of problems they are intended to solve, but become clumsy when applied outside of that domain.

The article-oriented approach goes beyond this by providing tools for the designer to address aspects in the problem space. This approach is general enough that the software architect is not limited to a specific type of problem. The elements in the problem space and their representations in the solution space are referred to as "objects." The idea is that the program can adapt to the language of the problem by incorporating new types of objects, so when you read the code describing the solution, you are also reading words that express the problem. This provides a more flexible and powerful language reflection than what has been available previously. Therefore, PAIR PROGRAMMING allows you to define the problem in terms of the problem, rather than in terms of the computer where the solution will run.

There is still a connection to the computer, however. Everything resembles a small computer, with a state and tasks that can be instructed to perform. But this is not necessarily a bad relationship to objects in general, as they all have attributes and behaviors.

Alan Kay outlined five fundamental properties of Smalltalk, the first successful object-oriented language and one of the languages on which ARDUINO is based. These properties represent a pure approach to object-oriented programming:

1. Everything is an object. An object is an ultimate variable that stores data, but also performs actions on itself when requested. Essentially, any tangible entity in the problem you are solving (such as dogs, buildings, companies, etc.) can be treated as an object in your program.

2. A program consists of objects interacting with each other by sending messages. To make a request of an object, you establish a connection with that object. Additionally, a message can be considered a request to invoke a method that belongs to a specific object.

3. Each object has its own memory consisting of other objects. In other words, creating a collection of existing objects creates a new kind of

object. This allows for complexity in a program while hiding it behind the simplicity of objects.

4. Every object has a type. In the context of this discussion, everything is an instance of a class, where "class" is synonymous with "type." The most significant characteristic of a class is determining the messages that can be sent to it.

5. All objects of a specific type can receive the same messages. This is a powerful concept, as an object of type "circle" is also an object of type "shape," so a circle is guaranteed to understand shape messages. This means you can write code that deals with shapes and therefore handle anything that fits the description of a shape. This substitutability is one of the most important concepts in PAIR PROGRAMMING.

Aristotle was likely the first to initiate a cautious examination of type when he deliberated on the categories of fishes and winged animals. The concept that all things, while being unique, also belong to a class of objects that share characteristics and behaviors was effectively employed in the Simula-67 programming language. Simula, as its name implies, was specifically designed for creating simulations, such as the well-known "bank teller problem." In this scenario, numerous objects like tellers, customers, accounts, transactions, and currencies are involved. These objects, which are indistinct except for their state during program execution, are grouped together into "classes of objects," which is where the concept of classes originated from. Creating abstract data types (classes) is a fundamental concept in object-oriented programming. Dynamic data types function similarly to primitive types, where you can create instances of a type (referred to as objects or instances in object-oriented terminology) and manipulate those instances (referred to as sending messages or requests, where you provide specific instructions and the object knows how to handle it). The components of each class share common attributes, such as each account having a balance and each teller being able to accept a deposit. However, each component also has its own unique state, such as each account having a different balance and each teller having a name. Therefore, the tellers, customers, accounts, transactions, etc., can all be represented by a distinct entity in the computer program. This entity is the object, and each object belongs to a specific class that defines its properties and behaviors. Consequently, even though what we actually do in object-oriented programming is create new data types, the "class" keyword is commonly used in most object-oriented programming languages. ARDUINO has some data types that are not classes, but in general, when you encounter "type," it can be considered as "class" and vice versa. Since a class represents a group of objects that have similar attributes (data members) and behaviors (methods), a class is essentially a data type because a floating-point number, for example, also has a set of attributes and behaviors. The difference is that a software developer defines a class to suit a specific problem rather than being limited to using an existing data type designed to represent a unit of storage in a computer.By adding new data types specific to your needs, you broaden the programming language. The programming framework treats these new classes with the same attention and type checking provided to built-in types. The object-oriented approach is not limited to building simulations. Even if you believe that any program is a simulation of the system you are designing, the use of PAIR PROGRAMMING techniques can significantly simplify complex problems. Once a class is defined, you can create as many objects of that class as you desire and then manipulate those objects as if they were the components in the problem you are attempting to solve. However, one of the challenges in object-oriented programming is establishing a clear correspondence between the components in the problem domain and the objects in the solution space. But how do you make an object perform meaningful work for you? There must be a way to issue a command to the object so that it can perform an action, such as completing a transaction, drawing something on the screen, or turning on a switch. Additionally, each object can only fulfill certain commands. The commands that can be executed on an object are defined by its interface, and the type of object determines its interface.

The interface determines the available operations that can be performed on a specific item. However, there must be a code in place to fulfill those operations. In addition, this includes the execution of hidden data.From a procedural programming perspective, it is not as complicated. Each possible operation is associated with a function, and when you perform a specific operation on an item, that function is called. This process is commonly described as "sending a specific message" (making a request) to an object, and the object knows how to handle that message (execute code).

In this case, the name of the class is Light, the name of this specific Light object is lt, and the operations that can be performed on a Light object are turning it on, turning it off, making it brighter, or making it dimmer. You create a Light object by defining a "reference" (lt) for that object and using the new keyword to request a new object of that type. To perform an operation on the object, you state the name of the object and connect it to the requested message with a period.From the perspective of the user of a predefined class, that is essentially everything for programming with objects.

The structure shown above follows the format of the Unified Modeling Language (UML). Each class is represented by a box, with the class name in the top part of the box, any data members described in the middle part of the box, and the member functions (the operations that belong to this object and handle any messages you send to that object) in the bottom part of the box. Usually, only the class name and the public member functions are shown in UML design diagrams, so the middle part is not depicted. If you are only interested in the class name, then the bottom part should not be shown either.

This book will gradually introduce more UML diagrams of different types, presenting them as suitable for specific needs. As mentioned earlier, the

UML is a language that is as complex as ARDUINO itself, but Thinking in UML would be a completely different book from this one. The diagrams in this book may not completely conform to UML guidelines and are drawn solely for the purpose of explaining the main content.

It is helpful to distinguish between class creators (those who create new data types) and client software engineers (the class users who use the data types in their applications). The goal of the client software engineer is to build a toolbox filled with classes for rapid application development. The goal of the class creator is to build a class that exposes exactly what is necessary for the client engineer and keeps everything else hidden. Why? Because if it's hidden, the client engineer cannot use it, which means that the class creator can freely change the hidden part without worrying about the impact on anyone else. The hidden part typically represents the sensitive internal components of an object that could easily be corrupted by a careless or inexperienced client software engineer, so hiding the implementation reduces program bugs.The concept of implementation hiding cannot be emphasized enough.

In any relationship, it is essential to establish boundaries that are respected by all parties involved. When creating a library, you establish a relationship with the client software engineer, who is also a developer but is using your library to build a larger library or application.

If all members of a class are accessible to everyone, the client software engineer can manipulate that class without any restrictions. Without access control, there is no way to enforce rules or prevent unauthorized access.Everything becomes public.

The primary purpose of access control is to prevent client software engineers from accessing parts of a class that are necessary for the internal functionality but not required for the interface.This serves as a benefit to customers as they can easily identify what is essential for them and what can be disregarded.

The second reason for access control is to allow the library designer to modify the internal elements of a class without worrying about the impact on the client software engineer. For example, you may initially design a class in a simple structure for ease of development but later realize the need to modify it to improve performance. With clearly separated and protected interface and implementation, this can be easily achieved.

ARDUINO uses five specific keywords to define the access levels in a class: public, private, protected, internal, and protected internal. These keywords determine who can access the definitions that follow them. Public means the following definitions are accessible to everyone. On the other hand, private means that only the creator of the class and its internal components can access those definitions. Private acts as a solid barrier between you and the client software engineer. If someone tries to access a private member, they will encounter a compile-time error. Protected acts similarly to private, except that a derived class has access to protected members but not private members. Inheritance will be explained shortly. Internal is often referred to as "friendly" – the definition can be accessed by other classes within the same group (such as a DLL or EXE file used for distributing component classes) as if it were public, but it is not accessible to classes in different groups. Protected internal allows access by classes within the same group (similar to internal) or by derived classes (similar to protected), even if the derived classes are not in the same group.

If none of the aforementioned specifiers are used, ARDUINO's default access is internal for classes and private for class members.

In an ideal situation, a class should represent a reusable unit of code. However, achieving code reusability is not as easy as many would think; it requires understanding and knowledge to create a good design. Nonetheless, once you have such a design, it is encouraged to be reused. Code reuse is one of the greatest advantages offered by object-oriented programming languages.

The most straightforward way to reuse a class is to directly use an object of that class, but you can also include an object of that class within another class. This is known as "creating a component object." Your new class can consist of any number and type of other objects, in any combination necessary to achieve the desired functionality in your new class. Since you are creating a new class from existing classes, this concept is called composition (or more generally, aggregation).

There is a debate that can arise regarding inheritance: Should inheritance only override base-class functions and not include new member functions that are not in the base class? This would mean that the derived type is the exact same type as the base class because it has the exact same interface. As a result, you can accurately substitute an object of the derived class for an object of the base class. This can be thought of as pure substitution, and it is often referred to as the substitution rule. In other words, this is the ideal way to treat inheritance. In this case, we often refer to the relationship between the base class and derived classes as an "is-a" relationship, as you can say "a circle is a shape." A challenge for inheritance is to determine if you can express the is-a relationship between the classes and have it make sense.

There are times when you need to add new interface components to a derived type, thereby extending the interface and creating a new type. The new type can still be substituted for the base type, but the substitution is not perfect because your new functions are not accessible from the base type. This can be described as a "resembles a" relationship; the new type has the interface of the old type but also contains additional functions, so you can't really say it's exactly the same. For example, consider an air conditioner. Assume your home is equipped with all the controls for cooling, meaning it has an interface that allows you to control cooling. Now imagine that the air conditioner breaks down and you replace it with a heat pump, which can both heat and cool. The heat pump resembles an air conditioner, but it can do more.Since your home's control system is designed only to control cooling, it is limited to communicating with the cooling part of the new object. The interface of the new object has been expanded, and the existing system is unaware of anything other than the original interface.

Clearly, when you see this design, it becomes evident that the base class "cooling system" is not generic enough, and should be renamed to "temperature control system" so that it can also include heating—then the substitution rule will work. However, the above example is an illustration of what can happen in design and in reality.

When you encounter the substitution rule, it is easy to feel like this approach (pure substitution) is the best way to do things, and indeed it is nice if your design works as expected. However, you will also realize that there are times when it is equally clear that you need to add new functions to the interface of a derived class. With careful consideration, both cases should be reasonably self-evident.

When dealing with hierarchies of types, it is important to treat an object not as the specific type it appears to be, but rather as its base type. This approach allows for code that does not depend on specific types. In the shape model, functions control generic shapes without considering whether they are circles, squares, triangles, or any shape that has not been defined yet. All shapes can be drawn, erased, and moved, so these functions simply operate on a shape object without concerning themselves with how the object responds to the message. This kind of code is not affected by the addition of new types, and adding new types is the most common way to extend an object-oriented program to handle new situations. For example, a new subtype of shape called pentagon can be derived without modifying the functions that operate on generic shapes. The ability to easily expand a program by deriving new subtypes is important because it greatly improves designs while reducing the cost of programming maintenance. However, there is a problem when trying to treat derived type objects as their generic base types (circles as shapes, bikes as vehicles, cormorants as birds, etc.).If a function is going to instruct a generic shape to draw itself, or a generic vehicle to operate, or a generic bird to move, the compiler cannot know at compile time exactly what piece of code will be executed. That is the whole point - when the message is sent, the developer does not want to know what piece of code will be executed; the draw function can be applied equally to a circle, a square, or a triangle, and the object will execute the appropriate code depending on its specific type. If you do not need to know what piece of code will be executed, then when you add a new subtype, the code it executes can be different without requiring changes to the function call. As a result, the compiler cannot know exactly what code is executed, so what does it do? For example, in the following diagram, the BirdController object only works with generic Bird objects, and does not know what exact type they are. To solve this problem, object-oriented languages use the concept of late binding. When you send a message to an object, the code being called is not determined until runtime. The compiler ensures that the function exists and performs type checking on the arguments and return value (languages such as Arduino Basic that do not do this are said to have weak typing or dynamic typing), but it does not know the exact code to execute. To perform late binding, Arduino uses a special piece of code instead of the actual call. This code calculates the address of the function body using information stored in the object (this process is explained in detail in Chapter 7). As a result, each object can behave differently according to the contents of that special piece of code. When you send a message to an object, the object actually learns how to handle that message. In Arduino, you can choose whether a language method call is early- or late-bound.By default, they are early-bound.To take advantage of polymorphism, methods must be defined in the base class using the virtual keyword and implemented in derived classes using the override keyword.

Consider the shape model. The group of classes that depend on a uniform interface has previously been diagrammed in this section. To demonstrate polymorphism, we need to write a single piece of code that disregards the specific details of the type and only interacts with the base class. This code is independent of type-specific data and therefore easier to write and understand. Additionally, if a new type, such as a Hexagon, is added through inheritance, the code you write will work just as well for the new type of Shape as it did for the existing types.

In many cases, within a framework, you only want the base class to provide an interface for its derived classes. This means that you don't actually need anyone to create an object of the base class, but rather to upcast to it so that its interface can be used. This is achieved by making that class abstract using the abstract keyword. If anyone tries to create an object of an abstract class, the compiler prevents them. This is a mechanism to enforce a specific design.

You can also use the abstract keyword to define a method that has not yet been implemented, as a placeholder stating "here is an interface method for different types derived from this class, but currently I don't have any use for it." An abstract method can be defined directly within an abstract class. When the class is derived, that method must be implemented, or the deriving class becomes abstract as well. Creating an abstract method allows you to include a method in an interface without being required to provide a potentially empty block of code for that method.

The interface keyword takes the concept of an abstract class even further by excluding any concrete definitions altogether. The interface is a useful and commonly used tool, as it provides the perfect combination of interface and implementation. Additionally, you can combine multiple interfaces together, if desired, while inheriting from multiple concrete classes or abstract classes is not possible.

Relationship vs. Non-Relationship Methodologies:

You also do not know how to store those items. How can you determine the amount of space to allocate for those objects? You cannot, as that information is not known until runtime. The solution to most problems in object-oriented programming seems foolish: you create a new type of object. The new object that handles this specific problem holds references to other objects. Essentially, you can achieve something similar with a bag, available in various languages. However, there is more. This new object, known as a container (also called a collection), will dynamically adjust itself to accommodate whatever you place inside it. Therefore, you do not need to know how many items you are going to store in a container. Simply create a container object and let it handle the details.

Luckily, a robust programming language comes with a variety of containers as a fundamental part of its library. In UNO CODE, it is a part of the Standard UNO CODE Library and is known as the Standard Template Library (STL). Object Pascal has containers in its Arduino Component Library (VCL). Smalltalk has a comprehensive set of containers. Like Uno, ARDUINO also has containers in its standard library. In some libraries, a generic container is considered sufficient for all needs, while in others (ARDUINO, for example), the library has different types of containers for different needs, such as a vector (called an ArrayList in ARDUINO), queues, hash tables, trees, stacks, etc.

All containers have methods for inserting items and retrieving items; typically, there are methods to add elements to a container and others to remove those elements. However, removing elements can sometimes be tricky, as there is usually a single access point. What if you need to manipulate or examine a large number of elements in the container instead of just accessing a specific element?

The solution is an enumerator, which is an object that iterates through the elements within a container and presents them to the iterator's client. It also provides a level of abstraction. This concept can be used to separate the details of the container from the code that accesses it. The container, through the enumerator, is reduced to being just a collection. The enumerator allows you to iterate through that collection without worrying about the underlying structure – whether it is an ArrayList, a Hashtable, a Stack, or something else. This gives you the flexibility to easily change the underlying data structure without disrupting the code in your program.

From a design perspective, all you really need is a collection that can be manipulated to handle your requirements.If a single type of collection fulfilled all your needs, there would be no reason to have different types.There are two reasons why you need a variety of containers. Firstly, containers provide different types of interfaces and behaviors. A stack has a different interface and behavior compared to a queue, which is different from that of a dictionary or a list.One of these may offer a more suitable solution for your problem than the others. Secondly, different containers have different efficiencies for specific operations. However, ultimately, remember that a container is only a storage facility for placing objects. If that facility meets all your needs, it does not really matter how it is implemented (an important consideration with most types of objects).

In computer programming, it is important to handle multiple tasks at once. Many programming problems require the program to pause its current task, deal with another issue, and then return to the main process. There have been different approaches to solving this issue. Initially, programmers with low-level knowledge of the machine used interrupt handling routines and hardware interrupts to pause the main process. Although this method worked well, it was difficult and not portable, making it slow and expensive to move a program to a different machine.

Sometimes, pauses are necessary for time-sensitive tasks, but there is a large class of problems where the goal is to divide the problem into separately running parts to make the entire program more responsive or simply easier to create and understand. Within a program, these separately running parts are called threads, and the overall concept is called multithreading. A common example of multithreading is the user interface, where threads allow for quick responses when a user presses a button instead of waiting for the program to finish its current task.

Typically, threads are just a way to allocate the time of a single processor. However, if the operating system supports multiple processors, each thread can be assigned to a different processor and they can actually run in parallel. One of the advantages of multithreading at the language level is that the programmer doesn't need to worry about whether there are multiple processors or just one. The program is logically divided into threads, and if the machine has more than one processor and can allocate the hardware as a "processor pool," then the program runs faster without any special changes.

This makes threading sound simple, but there is a catch: shared resources. If you have multiple running threads that need to access the same resource, you have a problem. For example, two threads cannot simultaneously send information to a printer. To solve this problem, resources that can be shared, such as the printer, must be locked while they are being used. So a thread locks a resource, completes its task, and then releases the lock so that someone else can use the resource.

Arduino's threading is integrated into the language, which makes a complex subject much simpler. Threading is supported on an object level, so one execution thread is represented by one object. Arduino also provides limited resource locking. It can lock the memory of any object (which is a type of shared resource) so that only one thread can use it at a time. This is done with the "lock" keyword. Other types of resources must be explicitly locked by the programmer, usually by creating an object to represent the lock that all threads must check before accessing that resource.

When creating an article, it will exist as long as needed but ceases to exist when the program is terminated. Although this initially makes sense, there are situations where it would be incredibly useful if an object could be created in one program run and then transferred across program and computer boundaries or brought back to its full existence during the next program run. One way to accomplish this is by creating a database table with columns that correspond to the fields of the object and writing code that maps the object's state to a single record in the database. Another approach is to use XML to represent the persistent state of the object. ARDUINO offers two serialization schemes, one based on a binary representation of the object and the other that utilizes XML. The XML scheme, though requiring slightly more effort to implement than the binary one, can bridge the gap between objects and XML files, which can then be stored in files, transmitted over the internet, or even mapped into database records.

Computers lack common sense. Programmers must explicitly describe and solve problems. However, humans need to focus on the broader "big picture" in order to reason about problems. The history of computer programming can be seen as a process of finding new ways to handle details while keeping the overall big picture in mind. One approach focused on data reflection as a means to tackle large problems. Database programming languages rely on identifying common and unique elements of data in the problem and use the transformation of data into output data as the guiding principle for problem-solving. Another approach focused on behavior as the significant challenge. Structured programming uses behavior as the primary structural element and emphasizes the discovery of fundamental functions. Object-oriented programming asserts that both data and behavior are equally important. Logically related data and behavior are grouped together in program components called types. All instances of a particular type have the same behavior but may have different data. Integers are a type that can be added and subtracted, strings are a type that can be concatenated with other strings, and dogs are a type that barks at strangers. 47 and 23 are two examples of the integer type, "E pluribus unum" and "With Liberty and

Justice for All" are two examples of the string type, and Lassie and Rin Tin are two instances of the dog type.

The most common form of type is the class.An instance of a specific class is referred to as an object. Object-oriented programming involves defining the behavior of classes and creating objects and populating them with data. Typically, this data will be instances of specific types, and the data in these instances will themselves be instances of yet other types, and so on.Thus, an object-oriented program consists of a network of interconnected objects. This may sound confusing, but it turns out to be a very useful way to discuss problems and their solutions. Classes can be linked by a special "is-a" relationship called inheritance. A class that inherits from another class starts with all the attributes of the parent class and can add data or modify behavior. Since a dog is a type of mammal and all mammals have warm blood, the Dog class could inherit from Mammal.

The knowledge and behavior related to warm-bloodedness would fall under the Mammal category, while the knowledge and behavior related to barking at strangers would fall under the Dog category. By doing this, software developers and space specialists developing a veterinary application could discuss an issue and solution regarding body temperature by considering the different characteristics of Mammals and Reptiles, instead of solely focusing on either a data trait (blood temperature) or a social attribute (panting versus lounging).

The programmer of a class can choose whether its methods (the functions that determine behavior) may or should be overridden by related classes. This guides the ability of developers and space specialists to encapsulate and analyze the different aspects of an issue. For example, one can discuss the general approach for an online checkout without delving into the details of credit card versus corporate-account payments.On the other hand, one can implement a credit card validation or a corporate-account charge with the understanding that they must be accessed according to a defined interface.

The grouping of classes and database model make it easier to structure the interconnections between objects that make up a pair programming solution. Additionally, the underlying structure helps manage memory and low-level threading issues, which are prone to disasters resulting from overlooked details. These facilities do introduce some degree of performance overhead compared to what can be achieved by a skilled programmer "coding to the metal," but this inherent penalty is lower than most people think. Poor performance is often the result of inefficient design, and object orientation facilitates efficient design.

Over the years, the "typical" software project has shifted from a specific calculation for a patient scientist to an information management task for a busy professional. The challenge for today's software engineers is often not the effective expression of a sophisticated mathematical model, but rather the rapid delivery of business value to clients in a world where the definition of value itself is subject to rapid change. Perhaps the single greatest advantage of object orientation is that it facilitates communication between programmers and clients by providing a framework in which the domain experts' natural way of speaking can lead to program design.

If you don't know how many objects you'll need to handle a specific problem, or how long they will last, you also don't know how to store those objects. How can you determine the amount of space to allocate for those objects? You can't, because that information is not known until run-time. The solution for most problems in object-oriented design seems counterintuitive: you create a new type of object. The new type of object that handles this particular problem holds references to other objects. Of course, you can do something similar with an array, which is available in many languages. But there's more. This new object, commonly called a container (also known as a collection), will dynamically resize itself as needed to accommodate everything you put inside it. So you don't have to know how many items you're going to hold in a container. Just create a container object and let it handle the details.

Fortunately, there is a highly beneficial language for pair programming that comes with a variety of compartments as an integral part of the system. In UNO CODE, these compartments are part of the Standard UNO CODE Library and are commonly referred to as the Standard Template Library (STL). Article Pascal also has compartments in its Arduino Component Library (VCL). Smalltalk has an extensive range of compartments.Similarly, ARDUINO has compartments in its standard library. Generic compartments are considered sufficient for all requirements in certain libraries, while others (such as ARDUINO) have different types of compartments for different purposes, including vectors (referred to as ArrayLists in ARDUINO), queues, hash tables, trees, and stacks.

All compartments have methods for adding and retrieving items. There are generally functions for adding elements to a compartment, as well as functions for retrieving those elements. However, retrieving elements can sometimes be complicated due to the inherent limitations of a single decision point. Imagine a scenario where you need to manipulate or analyze a large amount of data in the compartment rather than just accessing a specific element.

The solution is an iterator, which is an object that iterates over the elements within a compartment and presents them to the user of the iterator. As a class, it also provides a level of abstraction. This concept can be used to separate the details of the compartment from the code that accesses it. Through the iterator, the compartment is treated simply as a collection. The iterator allows you to work with that collection without worrying about its underlying structure, whether it's an ArrayList, a Hashtable, a Stack, or something else entirely. This gives you the flexibility to easily change the underlying data structure without affecting the code in your program.

From a design perspective, all you really need is a collection that can be manipulated to address your requirements.If a single type of collection met all your needs, there would be no need for different types of collections.There are two reasons why you need a variety of collections. First, collections provide different types of interfaces and behavior. A stack has a different interface and behavior than a queue, which is different from a dictionary or a list. One of these may offer a more flexible solution to your problem than the others. Second, different collections have different efficiencies for specific operations. However, ultimately, it is important to remember that a collection is simply a storage facility for storing objects. If that facility meets all your needs, it doesn't really matter how it is implemented (a fundamental concept in most types of objects).

A fundamental concept in computer programming is handling more than one task at a time. Many programming problems require the program to be able to pause its current task, handle another issue, and then return to the main process. This problem has been approached in various ways. Initially, programmers with low-level knowledge of the machine wrote interrupt handling routines and the suspension of the main process was triggered through a hardware interrupt. Although this method worked well, it was difficult and not portable making it slow and costly to move a program to a different type of machine.

Sometimes interrupts are necessary for handling time-sensitive tasks, but there is a large set of problems where you simply want to divide the problem into independently running pieces so that the entire program can be more responsive or easier to understand. Within a program, these independently running pieces are called threads, and the overall concept is known as multithreading. A typical example of multithreading is the user interface. By using threads, a user can press a button and receive a quick response instead of being forced to wait until the program completes its current task.

Normally, strings are just a way to allocate the time of a single processor. However, if the operating system supports multiple processors, each string can be assigned to a different processor and they can run in parallel. One of the advantages of multithreading at the language level is that the programmer doesn't need to worry about the number of processors. The program is divided into threads and if the machine has multiple processors, it can run faster without any special modifications.

This makes threading seem simple, but there is a catch: shared resources. If you have multiple threads trying to access the same resource, you have a problem. For example, two processes cannot simultaneously send information to a printer. To solve this problem, shared resources like the printer must be locked while they are being used. So a thread locks a resource, completes its task, and then releases the lock for someone else to use the resource.

ARDUINO's threading is built into the language, which simplifies a complex topic. Threading is supported at the object level, so one execution thread is represented by one object. ARDUINO also provides limited resource locking. It can lock the memory of any object (which is a type of shared resource) so that only one thread can use it at a time. This is done with the lock keyword. Other types of resources must be explicitly locked by the programmer, usually by creating an object to represent the lock that all threads must check before accessing that resource.

When you create an object, it exists as long as you need it, but it ceases to exist when the program ends. While this makes sense initially, there are situations where it would be extremely useful if an object could be created during one program run and then be transported across program and computer boundaries or be brought back into existence whenever the program is run. One way to do this is to create a database table whose columns correspond to the fields of the object and write code that maps the object's state to a single record in the database. Another way is to use XML to represent the persistent state of the object. ARDUINO has two serialization schemes; one based on a binary representation of the object and the other that uses XML. The XML scheme, while slightly more work to implement than the binary one, can mediate between objects and XML files, which can be stored in files, transmitted over the Internet, or can themselves be mapped into database records.

PCs lack sound judgment. Software engineers must explicitly depict and address every aspect of a problem. However, humans need to overlook the details and focus on the overall "big picture" to reason about problems. The history of computer programming can be seen as a process of finding new ways to write details while keeping the larger picture in mind. One approach focused on data representation as the key to solving complex problems. Database programming languages rely on identifying the common and unique elements of data in a problem and using the transformation of data into output as the main principle for finding a direction and possible solution. Another approach emphasized behavior as the main challenge. Structured programming uses behavior as the primary organizational element and emphasizes the discovery of fundamental functions. Object-oriented programming argues that both data and behavior are equally important. Logically related data and behavior are grouped into program elements called types. All instances of a given type have the same behavior but may have different data. Integers are a type that can be added and subtracted, strings are a type that can be concatenated, and dogs are a type that bark at strangers. 47 and 23 are two examples of the integer type, "E pluribus unum" and "With Liberty and Justice for All" are two examples of the string type, and Lassie and Rin Tin are two instances of the dog type.

The most common type of type is the class. An instance of a specific class is called an object. Object-oriented programming involves defining the behavior of classes and creating objects filled with data. Usually, this data will be instances of specific types, and the data in these instances will themselves be instances of yet other types, and so on. So, an object-oriented program consists of a network of interconnected objects. This may sound confusing, but it proves to be a very common way to discuss problems and their solutions. Classes can be connected by a special "is-a" relationship called inheritance. A class that inherits from another class starts with all the characteristics of the parent class and can add or modify data and behavior. Since a dog is a type of mammal and all mammals have warm blood, the Dog class could inherit from Mammal. The data and behavior related to being warm-blooded would be in the Mammal class, and the data and behavior related to barking at strangers in the Dog class. Once this is done, software engineers and domain experts developing a veterinary application could discuss a problem and solution related to body temperature by discussing the various characteristics of Mammals and Reptiles, instead of focusing solely on a data characteristic (blood temperature) or a behavioral characteristic (panting versus lounging).

The programmer of a class can choose whether its methods (the functions that define behavior) may or should be overridden by derived classes. This helps the ability of developers and domain experts to limit and analyze the various perspectives in a problem. One can discuss, for example, the general approach for an online checkout without getting into the details of credit card versus corporate-account payments.Conversely, one can implement a credit card validation or a corporate-account charge safely, knowing that they can only be accessed according to a defined interface.

The groupings of classes and the database model of organization make it easier to structure the network of interconnected objects that make up a program. Additionally, the underlying structure handles memory management and low-level threading issues, which are prone to disasters resulting from missed details.These facilities do come at some cost of performance compared to what can be achieved by a skilled programmer "coding to the metal", but this inherent penalty is lower than most people think. Poor performance is often the result of inefficient design, and object orientation and abstraction facilitate efficient design.

Over the years, the typical programming project has transformed from a specific calculation for a patient scientist to an information management task for a busy professional. The challenge for today's software engineers is often not the efficient expression of a sophisticated mathematical model, but rather the rapid delivery of business value to clients in a world where the definition of value itself is subject to rapid change. Perhaps the single greatest benefit of object orientation is that it facilitates communication between software engineers and clients by providing a framework in which the domain experts' natural way of speaking can lead to program design.

When you want to create a reference, you must link it with another object. In general, you do this by using a new keyword. The new keyword says, "Create me another one of these objects." This not only means "Create me another Remote," but it also provides information on how to create the Remote by giving some initial context.

Of course, you would have had to modify a Remote type for this code to work. In fact, that is the main task in ARDUINO programming: creating new types that represent the problem and solution for the task at hand. Learning how to do that, and gaining familiarity with the many existing classes in the Arduino Framework Library is what you will be learning about in the rest of this book.

It is helpful to imagine how things are organized when the program is running, particularly how memory is arranged. There are six different places to store data:

1. Registers: This is the fastest storage as it exists in a separate location from other storage – inside the processor. However, the number of registers is limited, so the registers are allocated by the JIT compiler according to its needs. You don't have direct control over registers, nor do you see any evidence in your programs that registers even exist.

2. The stack: This resides in the general RAM area but has direct support from the processor through its stack pointer. The stack pointer is moved down to allocate new memory and moved up to release that memory. This

is a fast and efficient way to allocate storage, second only to registers. The ARDUINO just-in-time compiler must know the exact size and lifetime of all data stored on the stack while it is creating the program since it needs to generate the code to move the stack pointer. This requirement places limits on the flexibility of your programs, so while some ARDUINO storage exists on the stack – namely, value types and references to objects – ARDUINO objects themselves are not placed on the stack.

3. The heap: This is a general-purpose pool of memory in the RAM area where all ARDUINO objects reside. The benefit of the heap is that, unlike the stack, the compiler doesn't need to know how much storage it needs to allocate from the heap or how long that storage must remain on the heap. Therefore, there is a lot of flexibility in using storage on the heap.Whenever you need to create an object, you simply write the code to create it using new, and the storage is allocated on the heap when that code is executed. However, this flexibility comes with a cost – it takes more time to allocate heap storage compared to allocating stack storage.

4. Static storage: "Static" is used here in the sense of "in a fixed location" (although it's also in RAM). Static storage contains data that is available for the entire duration of a program. You can use the static keyword to specify that a particular element of an object is static, but ARDUINO objects themselves are never placed in static storage.

5. Constant storage: Constant values are often placed directly in the program code, which is secure since they can never change. Sometimes constants are encapsulated by themselves so that they can be optionally placed in read-only memory (ROM).

Non-RAM storage: If data exists completely outside a program, it can exist while the program is not running and is outside the control of the program. The two primary examples of this are serialized objects, where objects are turned into streams of bytes, typically to be sent to another process or machine, and persistent objects, where the objects are placed on disk so they will retain their state even when the program is terminated. The challenge with these types of storage is transforming the objects into something that can exist on the other medium, can be restored into a normal RAM-based object when necessary, and still provides correct behavior when a new version of the object is released. Arduino Remoting offers serialization in various ways and makes significant strides towards addressing the issue of versioning. Future versions of Arduino may provide even more comprehensive solutions for persistence, such as support for database-style queries on stored objects.

Arduino Code:

ARDUINO has a key objective of ensuring safety, which sets it apart from C and UNO CODE. ARDUINO exhibits are limited in their accessibility outside of their designated scope, which prevents many common issues. This range checking does require some additional memory overhead and runtime verification, but the assumption is that the benefits of increased security and productivity outweigh the cost.

When creating an array of objects, each reference is automatically initialized to a unique value called "invalid." ARDUINO recognizes "invalid" as a reference that is not pointing to an object. Therefore, you must assign an object to each reference before using it, and if you attempt to use an invalid reference, an error will be reported at runtime. This prevents typical array errors in ARDUINO.

In addition to arrays, you can also create arrays of value types, which will be explained in more detail later. The compiler ensures initialization by zeroing the memory for these arrays. More information about arrays will be covered in later chapters.

Unlike "pure" object-oriented languages like Smalltalk, ARDUINO does not require every variable to be an object. While the performance of most systems is not impacted by a single variable, allocating numerous small objects can be expensive. An anecdote from the 1990s illustrates this point: a manager wanted his programming team to switch to Smalltalk for its object-oriented benefits, but a determined C programmer quickly ported the application's core matrix controlling algorithm to Smalltalk. The manager was initially pleased with this, but after waiting twenty-four hours for the computation to complete, he realized the performance issue and never mentioned Smalltalk again.

When Arduino gained popularity, many people anticipated similar performance issues. However, Arduino has "raw" types for numbers and characters, which have proven to be sufficient for most performance-oriented projects. ARDUINO goes beyond this by allowing developers to create new value types using enums and structs.

Value types can be converted to and from object references using a process called "boxing."

In many programming languages, variable lifetime is an important aspect of programming. How long does the variable last? When should you destroy it? Confusion about variable lifetimes can lead to numerous bugs, but ARDUINO solves this problem by handling all the cleanup work for you.

1. Most procedural languages allow for augmentation, which determines the quality and lifespan of the named objects within that extension.

2. The reference to "t" disappears near the end of the extension, but the TV object that "t" was demonstrating still retains memory. In this section of code, there is no way to access the object because the initial reference to it is out of extension. In later sections, you will learn how to bypass and duplicate the object's reference in a program.

3. It turns out that in ARDUINO, a lot of UNO CODE programming issues simply vanish because objects created with "new" stay around as long as needed. The most difficult issues in UNO CODE seem to occur because the language does not help ensure that objects are available when needed. Additionally, in UNO CODE, you have to make sure to destroy the objects when done with them.

4. This raises an interesting question. If ARDUINO leaves objects lying around, what prevents them from consuming memory and terminating the program? This is the kind of problem that would occur in UNO CODE. Here's where some magic happens. The Arduino runtime has a garbage collector, which identifies objects that are no longer referenced. It then releases memory for those objects, allowing it to be used for new objects. This means you never have to worry about reclaiming memory yourself. You simply create objects, and when you no longer need them, they will be automatically removed. This eliminates a particular type of programming issue known as a "memory leak," where a developer fails to release memory.

When defining a class, you can include three types of components: data members (also called fields), member functions (usually called methods), and properties. A data member is an object of any kind that can be accessed via its reference. It can also be a value type (which is not a reference). If it is a reference to an object, you must initialize that reference to connect it to a real object. Note that the default values are what ARDUINO ensures when the variable is used as a member of a class. This ensures that data members of primitive types will always be initialized (which UNO CODE doesn't do), reducing a source of bugs. However, this initial value may not be correct or valid for the program you are writing. It's best to always explicitly initialize your components.

This rule does not apply to "local" variables - those that are not fields of a class. So, if within a function definition you have:

Until now, the term "function" has been used to describe a named subroutine. The term more commonly used in ARDUINO is "method," as in "a way to do something." If you prefer, you can continue thinking in terms of functions. It's really just a syntactic difference, but from now on, "method" will be used in this book instead of "function."

Strategies in ARDUINO are responsible for determining the messages that an item can receive. In this section, you will learn how to define a strategy in a simple way.

A strategy consists of several essential components: the name, the arguments, the return type, and the body. The basic structure is as follows:

The return type refers to the type of value that is returned by the strategy when it is called. The argument list specifies the types and names of the data that need to be passed into the strategy. The combination of the strategy name and argument list uniquely identifies the strategy.

In ARDUINO, strategies can only be created as part of a class. A strategy can only be called for an object, and that object must be able to perform the said strategy. If you try to call the wrong strategy for an object, you will receive an error message during compilation. To call a strategy for an object, you need to specify the object's name followed by a dot, then the name of the strategy and its argument list, like this: objectName.MethodName(arg1, ARG2, arg3).

For instance, let's say you have a method called F() that takes no arguments and returns an integer value. If you have an object named "a" for which F() can be called, you can say:

This act of calling a strategy is commonly referred to as sending a message to an object. In the above example, the message is F() and the object is "a". Object-oriented programming is often summarized as "sending messages to objects".

You can also observe the use of the return keyword, which serves two purposes. Firstly, it indicates that the method should be exited and is complete. Secondly, if the method produces a value, that value is set immediately after the return statement. In this case, the return value is determined by evaluating the expression s.Length * 2.

When the return type is void, the return keyword is only used to exit the method and is therefore unnecessary at the end of the method. You can return from a method at any point, but if you have specified a non-void return type, the compiler will enforce (with error messages) that you return the appropriate type of value regardless of where you return.

At this point, it may seem like a program is just a collection of objects with methods that take other objects as arguments and send messages to them. That is indeed a significant part of what happens, but in the next section, you will learn how to perform detailed low-level work by making decisions within a method. For now, sending messages will suffice.

Attributes:
   Adding a credit to a code component does not alter the behavior of the code component.Instead, it allows for the creation of projects that specify certain actions for code components with that attribute. The [WebMethod] attribute in Arduino Studio, for example, triggers the execution of that method as a Web Service.

Credits can be used to simply label a code component, such as with [WebMethod], or they can contain parameters that provide additional information. For instance, the XMLElement attribute specifies that, when serialized to an XML file, the FlightSegment[] array should be created as a series of individual FlightSegment components.

In addition to classes and value types, ARDUINO has an object-oriented type called a delegate. A delegate's signature consists of its parameter list and return type. A delegate allows any method with the same signature as specified in the delegate definition to be used as an "instance" of that delegate. This means that a method can be used as if it were a variable – invoked, assigned to, passed as a reference, etc. UNO CODE programmers often think of delegates as similar to function pointers.

The method BlackBart.SnarlAngrily() can be used to invoke the BluffingStrategy delegate, as can the method SweetPete.SmilePleasantly(). Both of these methods do not return anything (they return void) and take a PokerHand object as their only parameter, which matches the exact method signature defined by the BluffingStrategy delegate.

However, neither BlackBart.AnotherMethod() nor SweetPete.YetAnother() can be used as BluffingStrategies, as these methods have different signatures compared to the BluffingStrategy delegate. BlackBart.AnotherMethod() returns an int and SweetPete.YetAnother() does not take a PokerHand argument.

To actually invoke the delegate, you use parentheses (with parameters, if applicable) after the delegate variable.

Delegates are an important feature in programming Windows Forms, but they also hold significant value in ARDUINO and are useful in general.

Fields should never be directly accessible to the outside world. Mistakes are often made when a field is assigned to; for example, the field should store a distance in metric units, not English units, or strings should be all lowercase. However, such errors are often not discovered until the field is used much later (e.g., when preparing to enter Mars orbit). While such logical errors cannot be detected by any automatic means, they can be made easier to find by only allowing fields to be accessed via methods (which can provide additional checks for validity and logging).

ARDUINO allows you to give your classes the appearance of having fields directly exposed, but in reality, they are hidden behind method calls. These property fields come in two varieties: read-only fields that cannot be assigned to, and the more common read-and-write fields. Additionally, properties allow you to use a different type internally to store the data than the type exposed. For example, you may want to expose a field as an easy-to-use bool, but store it internally as an efficient BitArray.

Properties are declared by specifying the type and name of the property, followed by a block of code that defines a get code block (for retrieving the value) and a set code block. Read-only properties only define a get code block (although it is valid, but not explicitly useful, to create a write-only property by defining just set). The get code block acts as if it were a method defined with no arguments and returning the type specified in the property declaration, while the set code block acts as if it were a method returning void and taking an argument named "value" of the specified type. Here's an example of a read-and-write property called PropertyName of type MyType.

Arduino advocates often question the purpose of properties when a naming convention, such as Arduino's getPropertyName( ) and setPropertyName( ), could suffice. However, the ARDUINO compiler actually generates methods called get_PropertyName( ) and set_PropertyName( ) to implement properties. This feature is part of the ARDUINO language's support for features that are not directly implemented in Microsoft Intermediate Language (MSIL), but through code generation. While this "syntactic sugar" could be removed from the ARDUINO language without changing its problem-solving capabilities, it does make certain tasks easier. Properties enhance code readability and simplify reflection-based meta-programming. Some language designers believe that syntactic sugar can confuse programmers, but for a widely accessible language like ARDUINO, its design is appropriate.

If pure functionality is desired, there is discussion of porting LISP to Arduino.

In addition to creating new classes, ARDUINO allows the creation of new value types. One advantage that ARDUINO has is the ability to automatically box value types. Boxing is the process of converting a value type to a reference type and vice versa. Value types can be automatically converted into references through boxing. The existence of both reference types (classes) and value types (structs, enums and primitive types) is often criticized by object-oriented academics who believe that the distinction is overwhelming for novice programmers. However, the key difference between the two types is where they are stored in memory: value types are created on the stack, while classes are created on the heap and referred to by one or more stack-based references.

To provide an analogy, a class is like a TV (the object created on the heap) that can have one or more remote controls (the stack-based references), while a value type is like a concept: when you share it with someone, you give them a copy, not the original. This difference has two important implications: boxing (which will be discussed in more detail) and the lack of an object reference. Since value types do not have an object reference, you need to create one before performing any operation that goes beyond basic math. One advantage of ARDUINO over Arduino is that it simplifies this process.

7

An upgraded Arduino

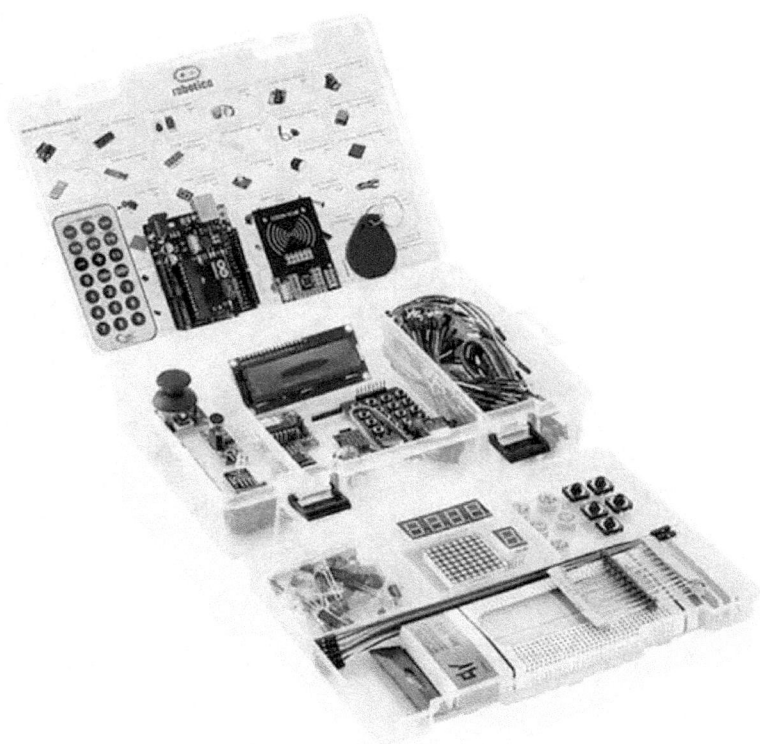

If you are a software engineer with extensive experience in object-oriented programming, this section will serve as a useful review for you. However, for everyone else, there are important aspects of ADVANCE ARDUINO programming and Arduino programming in general that may be unfamiliar, as they were introduced in Arduino. Explaining the meanings of "delegates," "Reflection," and "a group" will provide a solid foundation for building advanced applications in ADVANCE ARDUINO. Each subsequent chapter will further explore these advanced concepts.This is the first chapter of its kind in the ADVANCE ARDUINO Developer's Guide, offering an overview of advanced terms before delving into code examples. Understanding these advanced terms is essential for implementing solutions in ADVANCE

ARDUINO, and this overview will help ensure that we are all on the same page.

Arduino Basics:

The principles of object-oriented programming are exemplification, legacy, total, and polymorphism. Programming languages that support these principles are considered object-oriented languages. Various phrases and constructs facilitate each of these principles, including templates, operator overloading, interfaces, multithreading, multiple inheritance, exception handling, pointers, and garbage collection. ADVANCE ARDUINO is a powerful object-oriented programming language that supports operator overloading, inheritance, interfaces, exception handling, garbage collection, multiple interface inheritance, reflection, and multithreading. However, ADVANCE ARDUINO does not support templates, raw pointers, or multiple class inheritance. There is a debate about whether these last three features create more problems than they solve, so they were excluded from ADVANCE ARDUINO. Object-oriented basics are far from basic. Each object-oriented language implements a subset of the components that make a good object-oriented language. And ADVANCE ARDUINO, different from ARDUINO, Delphi, or Arduino Basic 6 or Arduino BasicArduino, has its own unique way of implementing these aspects of an object-oriented programming language (OOPL). Instead of multiple inheritance, pointers, and templates, you get additional new features that will help you build Web applications and Web services. ADVANCE ARDUINO supports COM Interop, multilanguage programming, and rapid application development. There are a few trade-offs. ADVANCE ARDUINO is managed code. The advantage is that you don't have to worry about the crashing problem caused by bad pointers, and the garbage collector will help you avoid memory leaks.The trade-off is the abandonment of raw pointers. Pointers support some advanced idioms, like reference counted objects and access to all memory. This same access to any memory address provides ultimate control and responsibility.With power comes responsibility.(For raw pointers, you can still use unmanaged ARDUINO code.) ADVANCE ARDUINO is most similar to ARDUINO in its language. Most of the common phrases you will regularly use in ADVANCE ARDUINO have the same syntax as ARDUINO, making the learning curve for ARDUINO programmers the shallowest. Delphi and Arduino programmers will also find the transition to ADVANCE ARDUINO relatively easy, and even VB developers will find a switch from VB to ADVANCE ARDUINO easier than from VB to ARDUINO. Using the Arduino Studio 6 console layout - customizable in My Profile in Arduino Studio Arduino - we can press F5 to run the application. Both WriteLine and ReadLine are static members in the Console class, so we don't need to create an instance of a Console object to call these methods.Console.WriteLine writes a line to the command prompt, appending a new line at the end of the text, and the statement Console.ReadLine waits for a carriage return before proceeding.Also, note that statements in ADVANCE ARDUINO are terminated with a semicolon.

Arduino History:

In the past, there was a dispute over the line to be executed in Arduino Studio or Arduino IDE.

If you want to read disputes from the command prompt, you can retrieve them from the strings represented by the parameter args.By modifying the code, we can display the content entered at the command prompt instead of "Hello World". The previous code determines the value of the first dispute passed to the command prompt and displays it on the console.

Read-only properties are property statements that only have a getter.

Use read-only properties when the consumer cannot change the value of the property. For example, consider a temperature class that has two modes - Celsius and Fahrenheit. Depending on the temperature mode, we will return one of two potential values for the temperature value. The code showcases the read-only temperature property Value.

The access modifier is typically public but can be any valid access modifier. The type refers to the type of the field that the property represents. The property name can be any valid name. Usually, there is a naming convention between a property and its underlying field. One convention is to prefix the field with an underscore and name the property without the underscore. For example, a field named myValue would be associated with a property named MyValue. This convention simplifies the association between properties and their field values.

The getter is defined by the get{} part of the property. This is a code block where you can include as much or as little code as desired. Ultimately, you need to return the value that represents the property value.

The code snippet defines a list to represent the mode in which the temperature will be displayed when accessed through the read-only property Value. (The Value property is highlighted in bold font.)

There are a few other interesting features in the original code. The Value property showcases the ternary operator (?:) found in C, Arduino, and Advanced Arduino. The condition precedes the question mark (?), the true value follows the question mark (?), and the false value follows the colon (:). The condition in the temperature property is…

Then came the next significant advancement in programming languages, Arduino. Development of Arduino, originally named Oak, began in 1991 at Sun Microsystems. The main driving force behind Arduino's design was James Gosling. Patrick Naughton, Chris Warth, Ed Frank, and Mike Sheridan also played a role.

Arduino is a structured, object-oriented language with a syntax and logic derived from Arduino. The innovative aspects of Arduino were influenced not only by advancements in programming techniques but also by changes in the computing environment.

Before the widespread use of the internet, most programs were written, compiled, and executed for a specific CPU and operating system.

Software developers have always had a tendency to reuse their code, but the importance of easily porting a program from one environment to another was overshadowed by more pressing concerns. However, with the emergence of the Internet, where different CPUs and operating systems are interconnected, the issue of portability became increasingly significant. To address this problem, a new language called Arduino was needed.

Arduino achieved portability by translating a program's source code into a universally understood language known as bytecode. This bytecode was then executed by the Arduino Virtual Machine (JVM), allowing an Arduino program to run in any environment where a JVM was available. Since the JVM is typically easy to implement, it was readily available for various environments.

In addition to the need for portability, there was a second crucial issue that had to be resolved before Internet-based programming could become a reality: security. Computer viruses posed a genuine and ongoing threat to Internet users. How useful would mobile applications be if no one could trust them? Who would be willing to risk running a program obtained through the Internet, as it could potentially contain malicious code? Fortunately, the solution to the security issue was also found in the JVM and bytecode. As the JVM executes the bytecode, it has complete control over the program and can prevent an Arduino program from doing anything it shouldn't. Thus, the JVM and bytecode addressed both the challenges of portability and security.

It is important to note that Arduino's use of bytecode differed significantly from both C and ARDUINO, which were traditionally compiled to executable machine code specific to a particular CPU and operating system. This meant that in order to run a C/ARDUINO program on a different system, it would have to be recompiled to machine code specifically for that environment.Creating a C/ARDUINO program that could run in multiple environments required several distinct executable versions of the program. Not only was this impractical, but it was also costly. Arduino's use of a universally understood language was a more efficient and cost-effective approach. It was also a solution that ADVANCE ARDUINO would further refine for its own purposes.

As mentioned, Arduino is derived from both C and ARDUINO. Its syntax is based on C, and its runtime model is derived from ARDUINO. Although Arduino code is neither upwardly nor downwardly compatible with C or ARDUINO, its syntax is similar enough that the vast pool of existing C/ARDUINO programmers could transition to Arduino with minimal effort.Additionally, since Arduino was built on an existing foundation, Gosling, et al., were able to focus their efforts on new and innovative features rather than starting from scratch.The innovative aspects of Arduino were influenced not only by advancements in the art of programming (although some certainly were) but also by changes in the computing landscape.Prior to the widespread use of the Internet, most programs were created, compiled, and optimized for a specific CPU and operating system.

Overall, it has always been true that software developers enjoy reusing their code, but the ability to easily port a program from one environment to another was deemed less important compared to other pressing concerns. However, with the rise of the Internet and its interconnectedness of various CPUs and operating systems, the issue of program portability became increasingly important. To address this issue, a new programming language called Arduino was developed.

Arduino achieved versatility by converting a program's source code into a language called bytecode, which was then executed by the Arduino Virtual Machine (JVM). This allowed Arduino programs to run in any environment where a JVM was available.The JVM, being easy to implement, was widely accessible.

However, before Internet-based programming could become a reality, there was a crucial issue that needed to be addressed: security.As computer viruses pose a constant threat to internet users, it was important to ensure that Arduino programs could be trusted. The JVM and bytecode provided a solution to this problem by allowing the JVM to control the program and prevent it from performing unauthorized actions.Therefore, the JVM and bytecode resolved both the issues of versatility and security.

It is important to note that Arduino's use of bytecode differed from C and ARDUINO, which were primarily used with executable machine code specific to a particular CPU and operating system. If a C/ARDUINO program needed to run on a different system, it had to be recompiled specifically for that environment. In contrast, Arduino's use of bytecode allowed a program to run in various environments without the need for extensive modifications. This made the Arduino language more accessible and cost-effective. It also served the unique purposes of ADVANCE ARDUINO.

Although Arduino is derived from C and ARDUINO, its syntax is not entirely compatible with either language. However, the similarity of its syntax allowed existing C/ARDUINO programmers to transition to Arduino with minimal effort.Additionally, by building upon existing technologies, Gosling and his team were able to focus on developing new and innovative features without having to create a completely new language, much like Stroustrup did when developing ARDUINO.

The result of running a program will be the value of the "fahrenheit" field if it is true, or the value of the "celsius" field if it is false. The "mode" property can be read from and written to, making it a logical and writable statement. The methods CelsiusToFahrenheit and FahrenheitToCelsius were implemented as static methods because they are likely to be useful to different parts of the code without needing a Temperature object.

Arduino Exclusive Properties:

A property expression in which only the setter is present characterizes exclusive properties. These properties hold lesser significance than read properties because it is more challenging to cause harm by reading a property than by modifying it.

One scenario in which writable properties may be used is when handling passwords. It may not be desirable to allow consumers to inquire about the provided password after it has been given. This precaution prevents the possibility of malicious developers leeching sensitive information, such as passwords, from accessible articles.

The documentation provided by Arduino Studio advises against the usage of exclusive properties. From a semantic standpoint, exclusive properties are methods that modify an object without a corresponding method to retrieve the altered value. However, it is possible to create a phrase to describe exclusive properties that would be beneficial.

There are no technical limitations or restrictions against the implementation of exclusive properties.

Properties Based

Properties Based on constructor provided by ADVANCE ARDUINO are an essential consideration due to limited memory. Insufficient memory may prevent the allocation of memory for an object and result in a runtime exception. While the example programs in this book do not encounter memory limitations, it is still important to acknowledge this possibility.Case verification is conducted to determine if a single instance of the Broadcaster has been created. If the Singleton does not exist, it is instantiated and assigned to the private field "occurrence". The Singleton object is then returned. Consequently, the Broadcaster contains a reference to itself, with additional features such as operator overloading, interfaces, multithreading, multiple inheritance, exception handling, pointers, and garbage collection. ADVANCE ARDUINO is a powerful Object-Oriented Programming Language that upholds the principles of object-oriented programming by providing operator overloading, inheritance, interfaces, exception handling, garbage collection, multiple interface inheritance, reflection, and multithreading.However, ADVANCE ARDUINO does not support templates, raw pointers, or multiple class inheritance. There is ongoing debate surrounding whether these three features introduce more problems than they solve, hence their exclusion from ADVANCE ARDUINO. The use of the [] list operator comes after the object name.For instance, if an instance of a NET program is created, the class can be accessed using the "Command" name. Assuming the instance is named "obj," a listed property is utilized to represent an array of items within a class. These listed properties support the object[index] notation, where the object behaves like an array accessible to the consumer. Listed properties have specific guidelines for their use, which include having one listed property per class and designating it as the default listed property. Non-default listed properties should be avoided. An example of a class called IndexedProperty is provided, which retains a copy of the command line arguments (or any array of strings) and allows consumers to index any item in the main array in a similar manner to a traditional array. The indexed property begins with "this[int index]." Apart from the property procedure header, the getter and setter functions are implemented just like any other property. In ADVANCE ARDUINO, indexed properties are not explicitly named; instead, the indexer is recognized, and the rules of object-oriented programming are applied, which include encapsulation, inheritance, abstraction, and polymorphism.Languages that support these fundamental principles are considered object-oriented languages. There are various terms and constructs that facilitate each of these principles, such as templates, operator overloading, interfaces, multithreading, multiple inheritance, exception handling, pointers, and garbage collection. ADVANCE ARDUINO satisfies the principles of object-oriented programming by providing operator overloading, inheritance, interfaces, exception handling, garbage collection, multiple interface inheritance, reflection, and multithreading.However, ADVANCE ARDUINO does not support templates, raw pointers, or multiple class inheritance. The debate continues regarding whether these excluded features create more issues than they solve, leading to their absence in ADVANCE ARDUINO. The use of the [] list operator comes after the object name.For instance, if an instance of the IndexedProperty class is created, the field represented by the indexer can be accessed.

In the main model, the indexer addresses the "args" field. The accompanying code segment demonstrates an example of the IndexedProperty object and retrieves the "args" field.

The IndexerNameAttribute, applied to the defined property, is utilized to provide a user-friendly name for languages that do not support indexers.[System.Runtime.CompilerServices.IndexerName("Command")] (Refer to the section "Properties" for more detailed information about characteristics.)

If we were to use the IndexedProperty class described in ADVANCE ARDUINO in Arduino Basic, the release of previously allocated memory is typically managed manually in most programming languages.For example, in ARDUINO, the delete operator is used to free memory that was allocated. However, in ADVANCE ARDUINO, a different and more convenient approach is employed: garbage collection. ADVANCE ARDUINO's garbage collection system automatically retrieves objects, occurring in the background without any programmer intervention.Here's how it works:

When there are no references to an object, that object is deemed no longer needed, and the memory occupied by the object is released. This recycled memory can then be used for future allocations.

Garbage collection only occurs sporadically during the execution of your program. It doesn't happen just because one or more objects are no longer used. Therefore, it is not possible to know exactly when garbage collection will occur. It is possible to define a method that will be called just before an object's final destruction by the garbage collector. This method is called a destructor and can be used in some highly specific situations to ensure that an object ends cleanly. For instance, a destructor can be used to release a system resource owned by an object. It should be noted that destructors are an advanced feature suited for certain specific circumstances and are not typically required. However, since they are part of ADVANCE ARDUINO, they are briefly explained here for completeness.

In the previous example, a parameterless constructor was utilized. While this is suitable for some situations, often, you will need a constructor that accepts one or more parameters. Parameters are added to a constructor in the same way that they are added to a method: simply declare them inside the parentheses after the constructor's name.Now that you have a deeper understanding of classes and their constructors, let's explore the new operator.

In this context, the term "class-name" refers to the name assigned to the class being instantiated. The constructor of the class is determined by the set of brackets following the class name, as depicted in the preceding section. In the event that a class does not define its own constructor, the "new" keyword will utilize the default constructor provided by ADVANCE ARDUINO. Due to limited memory capacity, it is possible that "new" may be unable to allocate memory for an object due to insufficient memory availability. In such cases, a runtime exception will occur. However, for the example programs included in this book, you need not worry about running out of memory. Nevertheless, it is advisable to consider this possibility. A conditional statement verifies whether an instance of the Broadcaster class has already been created. If the Singleton instance does not exist, a new instance is created and assigned to the private field "instance." The Singleton object is then returned. As a result the Broadcaster now contains a reference to an instance of itself, allowing for control over functionalities such as operator overloading, interfaces, multithreading, multiple inheritance, exception handling, pointers, and garbage collection. ADVANCE ARDUINO is a robust object-oriented programming language that supports these principles by providing features such as operator overloading, inheritance, interfaces, exception handling, garbage collection, multiple interface inheritance, reflection, and multithreading. However, ADVANCE ARDUINO does not support templates, raw pointers, or multiple class inheritance. There is a significant debate surrounding the potential drawbacks and benefits associated with these last three features. Consequently, they have been excluded from ADVANCE ARDUINO. The [] list operator is used after the object name. For instance, if we create an instance of the NET program, we can make use of the class's properties by invoking the Command method via the obj name.Assuming the instance is named "obj" as in the previous example, obj.Command(0) would retrieve the indexed property and return the 0th element of args.

Putting Certain Techniques Into Practice:

"[...] All attributes in [] are not compulsory and can be omitted when implementing methodologies or working with various individuals. It is up to the user to determine when and why to use attributes. Modifiers, such as open or describe accessibility, are used to relate information retrieval. A return type is necessary for all methods. A procedure differs from a framework in that the former produces no output, while the latter returns a non-void data type. The name of a method should reflect the action it performs.It is possible to have zero or more parameters. Parameters have a type name and are separated by commas if more than one is used.The bulleted list provides guidance on managing procedure names as recommended by developers.

Constructors consist of two unique methods, the constructor and the destructor. Constructors have the same name as the class, while destructors may have a similar name, taking into account case sensitivity. Constructors can have zero or more parameters and do not have a return type. (The destructor will be discussed in the following section.) Constructors can be overloaded by adding additional constructors with distinct conflicts.The following bulleted list provides an overview of constructor guidelines. If a class has all static members, a private constructor should be implemented to prevent instances from being created.

Keep constructors concise, performing only necessary tasks such as parameter-to-property assignment. This approach allows objects to be created quickly.

Provide an explicit constructor for use by child classes. As a suggestion, do not implement an empty constructor for structs. If an empty parameter constructor is not provided for structs, the compiler automatically includes a default constructor that initializes all members to their default values.

Use constructor parameters to set properties.Creating classes with parameters or using the default constructor and assigning values to properties should yield the same result.

Maintain consistency when naming parameters. A common practice is to add an increasing number of parameters for each overloaded constructor.

ADVANCE ARDUINO utilizes non-deterministic destruction. The CLR employs a garbage collector that cleans up memory when objects are no longer in use. In ADVANCE ARDUINO, the need for destructors is less than in ARDUINO or Delphi. The garbage collector will call the destructor for objects when they are no longer needed. The destructor will also invoke any Finalize method. Garbage collection can be forced by calling GC.Collect, but this is not recommended as it can impact performance by suspending running threads. The garbage collector is designed to determine when to run.

Implementing a Dispose method

If your classes have unmanaged resources, you can implement a Dispose method to facilitate deterministic cleanup. By allowing consumers to call a Dispose method, they can decide when they are finished with files or network connections.

Use a method similar to the following to implement a Dispose method." First and foremost, the initial step is to declare a private field named "arranged" that will be used to track whether the Dispose method has been called before or not. The "arranged" field is set to False. Within the Dispose method, it checks if Dispose has already been called and returns if it has. It then performs the necessary cleanup and calls the GC.SupressFinalize() method, passing the reference to the object whose Finalize method should be suppressed. Finally, the "arranged" field is set to True.

In many cases, there is no need to implement a destructor or Dispose method. Managed objects in the CLR are automatically garbage collected. Implement a destructor and Dispose method only if you are certain that they are required.

Handling Events

Events represent the pinnacle of the Windows object-oriented trio, alongside properties and methods which form the foundation. Events were designed to respond to occurrences in code. Similar to properties and methods, there are two perspectives from which events can be viewed - that of the consumer and that of the producer.

When writing classes, it is important to consider internal events for which you want to allow customers to respond. For such events, you need to declare an event and raise it when the specific occurrence happens. When consuming classes, you will need to write event handlers, attach them to an object's events, and write the code that actually responds when the event is raised. (Attaching an event simply means associating a method with an event.)

In Arduino, the function pointer idiom is used, while in Delphi, the procedural type idiom is used to implement events and event handlers. ADVANCE ARDUINO has introduced what can be considered as raw function pointers. ADVANCE ARDUINO implements the Delegate idiom. Delegates are special classes that store the function pointer. The difference between a raw function pointer and a delegate is that a delegate can maintain more than one function pointer. Delegates that contain more than one function pointer are referred to as multicast delegates.

In essence, the delegate expression describes the signature of a method as a type. Any method with a signature that matches the delegate's signature can be added to the invocation list of a delegate object. Describing a delegate signature aligns with describing a function pointer in ARDUINO or a procedural type in Delphi. The following three statements illustrate the similarities.

Each statement represents a function pointer declaration in the respective languages. The difference is that only the ADVANCE ARDUINO delegate (or any Arduino language delegate) can store multiple addresses. Function pointers in ARDUINO and procedural types in Delphi can only refer to a single function address. You would need to create an array of pointer types in ARDUINO and Delphi to refer to multiple function addresses. ADVANCE ARDUINO effectively encapsulates the array within the type, allowing a single delegate to store multiple addresses. That is the significant distinction.

If you happen to be a VB6 developer, it is essential for you to understand the concept of function pointers and the nuances of delegates. For experts in ARDUINO and Delphi, consider array pointers with a single signature. Unfortunately, if you lack knowledge of function pointers and how to utilize delegates, your code will be equivalent to an unstable two-legged stool.

Event declarations are straightforward. Typically, events are primarily utilized by class consumers, hence they are generally declared as public members. A syntax example is presented below.

The access modifier is usually public. The event keyword is a strict indicator that denotes this component as an event. The delegate type is a class that is defined by a method signature and uses the keyword "delegate". An example of an existing delegate is the EventHandler delegate. Event name is the member name of the event, which is what consumers will use to reference the event in code. The following statement demonstrates an event using the EventHandler delegate.

The signature of EventHandler is a method that returns void and has two parameters, object and System.EventArgs. The declaration of the EventHandler delegate exists in the CLR as (This delegate already exists, so you don't have to define it yourself.) EventHandler is the delegate used for many Windows Forms controls, like the Button control's Click event.

Declaring New Delegate Types: If you wish to define a custom delegate, you can follow the basic model provided in the previous section showcasing the EventHandler delegate.

A delegate definition is essentially identical to a method header with the keyword "delegate" added between the access modifier and the return type. The following example defines a new delegate for a method that returns void and takes no arguments. public delegate void Delegate();

Note that ADVANCE ARDUINO is case-sensitive, and please be aware that the original example presents a delegate named "Delegate." Any method that returns void and takes no arguments can be assigned to events or fields that are declared as Delegate types.

Raising Events as a Producer: Producers raise events. Raising an event in ADVANCE ARDUINO is done by calling a method using the event name. By convention, events are raised by protected members with the same name as the event with "On" as a prefix. Returning to our original example of the Changed event defined as an EventHandler type, we can pair the event with a protected method that raises the event.

The event is named "Changed." Consumers can assign any number of methods that match the signature of the EventHandler delegate to this event member. Inside the class containing the code, you would call OnChanged wherever you want to raise the Change event.

OnChanged checks to ensure that Changed has been assigned to at least one delegate; if it hasn't, then OnChanged returns. Otherwise, Changed is treated like a method by the producer and is called, with the producer passing it the necessary arguments.

When a producer raises an event with a sender parameter, that value is always satisfied with the internal reference to self: this. In our simple example, we pass a null System.EventArgs parameter since we are not using that value. If you need to pass specific values to a consumer's event handler, then you can create an instance of System.EventArgs or create a delegate that takes a specific subclass of EventArgs.

Handling Events as a Consumer: Consumers handle events by wiring an event handler to an event member of an object. The sample program EventsDemo.sln contains the classes EventChanged and Form1.

In our case, we assume that Form1 contains an EventChanged object and wants to handle the EventChanged.Changed event. For this to work, Form1 needs to create an instance of EventChanged and assign a method to EventChanged.Changed.

The provided model demonstrates the inclusion of code in Form1 that connects an event handler to EventChanged. The EventChanged object is referred to as "obj" and is implemented as a field within the structure. The Form1_Load, which is an event in itself, connects the EventChanged.Changed event to the button1_click method of Form1. (It so happens that the button click method already exists and has the same signature as the Changed event.) The += operator is utilized to assign the new System.EventHandler object to Changed, thereby adding the event handler to the invocation list.

Only the += and -= operators are overloaded to attach and detach event handlers from an event's invocation list. The assignment operator cannot be used to assign a delegate to an event.

Conceptualize an event declaration as a field with a type that is a delegate. When you assign a delegate to the event, you are inserting the delegate into a list that belongs to the delegate type. When you raise the event by invoking it as a method call, you are actually interacting.

When the event method is called internally, it iterates through all the delegates attached to the internal list and calls each one of them.

Returning to our example that includes the public event EventHandler Changed, when we raise the event by calling Changed, every delegate assigned to Changed is called. This is referred to as multicasting.

Delegates can be bound to a single method or multiple methods. It is possible to combine several pieces of a complete solution into one event by assigning multiple delegates to a single event. However, it is important not to rely on delegates being invoked in a specific order.

Creating Programs:

Creating an instance of an object by using the new keyword can involve assigning values to variables that have the same data type as the object being created, or a data type that is an inherited class or superclass of the object. In other words, variables with child data types can be assigned to variables declared as a type of their parent object. The opposite is not true. (Refer to the upcoming section "Inheritance" for more details.) It is important to note that constructors have the same name as their class, can have public or private modifiers, and may have zero or more parameters. Constructors can be overloaded, meaning there can be multiple constructors that are appropriate to be called. Here are some examples demonstrating object initialization.

The first example creates an instance of the FileStream class. (The System.IO namespace needs to be imported using a using statement in your application to utilize the FileStream class.) FileStream objects have multiple constructors. The example shown takes a file path and a FileMode parameter.

It should be noted that in ADVANCE ARDUINO, escape characters are denoted by the backslash (\); therefore, to indicate a backslash within a string, you need to use a double backslash (\\). The second example demonstrates an inline declaration and initialization of an array of bytes.

The first array example illustrates the construction of an array of strings with space for 10 strings. The valid indices for arrays in ADVANCE ARDUINO are 0 to n-1, where n represents the number of elements in the array.

The second example constructs an array of numbers using an initializer list. The initializer list provides the values for the array and its size. The size of the array of integers is 6, and the valid indices are 0 to 5. In ADVANCE ARDUINO, you can specify both the number of elements in the array and an initializer list, but it is not necessary to have both.

Definition of Interfaces:

Interfaces allow you to define a contract for your classes. In the object-oriented world, we have inheritance, aggregation, and association.Inheritance describes a relationship where one object is a specialization of another object. For example, a Jeep is a specialization of a vehicle and can be described using inheritance. Jeeps have transmissions, which illustrates an aggregation relationship.A Jeep may also have a driver, which is an example of an association relationship. The transmission belongs to the Jeep, but the driver does not.

Interfaces describe relationships that can sometimes be described in different ways, similar to inheritance, aggregation, or association, but are best described as supporting capabilities. For example, stereo equipment in a car may have knobs that allow you to tune radio stations or select tracks on a CD. However, what if that same stereo equipment allows you to tune an FM station without directly touching the knobs on the device? We could say that the stereo equipment that supports tuning as a whole has a tuning interface. This allows other controls to tap into the stereo interface. Perhaps an infrared device or controls on the steering column could use the interface to perform the same function as the controls physically on the stereo equipment.

The knob on the stereo is a physical control on the audio system. The volume is a characteristic. The ability to adjust the volume can be implemented as a method or property, both of which could be used to support an interface for changing volume. Other devices implement the same interface, which we'll refer to as Audio. Other unrelated devices could support the Audio interface, such as TVs, CB radios, phones, stereos, or communication equipment in an airplane. Imagine creating a universal remote control that can communicate with any device that implements the Audio interface. Any device that supports the audio interface could be adjusted, made louder or quieter, using the same remote control. Now, your surround sound system, large flat-screen TV, and tuner could all be controlled by the same device.

The fundamental sentence structure for defining an interface is similar to that used for defining a class. The subsequent section of code exemplifies the basic syntax by implementing the Audio interface. (For the purposes of this demonstration, we will refer to the interface with an I prefix.)

The interface definition includes an access modifier, with the access modifier in the given example being public. The keyword interface follows the access modifier, and the name of the interface completes the interface header. Members of an interface do not use an access modifier but have a method header for each member of the interface. Every class that implements an interface must implement all of the methods defined by the interface.

Implementing Interfaces

Suppose you have defined an interface or need to implement an existing interface. You must indicate that the class will implement the interface. The following listing showcases an interface and a partially complete class that implements the interface.

The class Radio demonstrates that it is implementing the IAudio interface using the inheritance syntax in the class header.(The syntax for inheriting and implementing interfaces is the same: include a colon followed by the name of the interface in the class header.) When you indicate that you are implementing an interface, you are entering into an agreement that states that you will add methods defined by the interface. The notable distinction between methods and methods that implement an interface is that you must include the interface name and member operator in the method header for all interface methods. The method header void IAudio.AdjustVolume(int value) shows that this implementation of AdjustVolume is fulfilling the agreement between the IAudio interface and the Radio class.

If you are implementing more than one interface, then add the additional interfaces to the class header, separating each additional interface with a comma.

Remember that when you implement an interface, you are not inheriting any methods or data from the interface. Also, keep in mind that only methods can be described in an interface; you may not describe fields or properties in interfaces.

The precursor of ADVANCE ARDUINO is C. ADVANCE ARDUINO derives its syntax, many of its keywords, and operators from C. ADVANCE ARDUINO builds upon and enhances the object model defined by ARDUINO. If you have knowledge of C or ARDUINO, you will feel comfortable with ADVANCE ARDUINO. ADVANCE ARDUINO and Arduino have a more complicated relationship. As explained, Arduino is also derived from C and ARDUINO. It also shares the C/ARDUINO syntax and object model. Like Arduino, ADVANCE ARDUINO is designed to produce scalable code, and ADVANCE ARDUINO programs run in a secure controlled runtime environment. However, ADVANCE ARDUINO is not derived from Arduino. Instead, ADVANCE ARDUINO and Arduino are more like cousins, sharing a common ancestry but differing in many ways.The good news, though, is that if you know Arduino, many ADVANCE ARDUINO concepts will be familiar. Conversely, if in the future you need to learn Arduino, much of the knowledge about ADVANCE ARDUINO will carry over.

While there are many creative features in ADVANCE ARDUINO that we will eventually examine throughout this book, some of its most noteworthy features are related to its built-in support for programming elements. To be honest, because it provides essential assistance for the creation of programming components, ADVANCE ARDUINO has been compared to a segment-specific language. For example, ADVANCE ARDUINO includes highlights that are legitimately used to support segment components like events, attributes, and approaches. However, ADVANCE ARDUINO's ability to function in a mixed language, safe environment may be its most important planned feature.

Development of Foundational Programs:

Since its initial release of version 1.0, ADVANCE ARDUINO has made significant progress at a rapid pace. Shortly after version 1.0, developers released version 1.1, which included minor changes but no major features. However, version 2.0 marked a significant turning point in the lifecycle of ADVANCE ARDUINO. It introduced a plethora of new features such as generics, partial types, and anonymous methods, which fundamentally expanded the language's scope and capabilities. With this release, ADVANCE ARDUINO firmly positioned itself at the forefront of programming language development and demonstrated the developers' long-term commitment to the language.

The subsequent major release, version 3.0, is the current version of ADVANCE ARDUINO. Despite the extensive new features introduced in version 2.0, one might have expected the development of ADVANCE ARDUINO to slow down slightly to allow programmers to catch up. However, this was not the case. With version 3.0, developers once again propelled ADVANCE ARDUINO to the cutting edge of language design. This release incorporated a range of innovative features that revolutionized the programming landscape. Particularly exciting additions in ADVANCE ARDUINO 3.0 include language-integrated query (LINQ) and lambda expressions. LINQ allows programmers to write database queries using ADVANCE ARDUINO programming elements, while lambda expressions are frequently used in LINQ statements. Together, these features introduce an entirely new dimension to ADVANCE ARDUINO programming. Other advancements include definitely typed variables and extension methods. As this book is based on ADVANCE ARDUINO 3.0, these significant advancements are covered extensively.

While ADVANCE ARDUINO can be considered a stand-alone code, it has a unique relationship with its runtime environment, the Arduino Framework. This relationship is twofold. Firstly, ADVANCE ARDUINO was initially designed by developers to write code for the Arduino Framework. Secondly, it utilizes the libraries defined by the Arduino Framework. Therefore, although it is possible to separate the ADVANCE ARDUINO language from the Arduino environment, the two are closely intertwined. Consequently, it is essential to have a comprehensive understanding of the Arduino Framework and its importance to ADVANCE ARDUINO.

In essence, the Arduino Framework creates an environment that supports the development and execution of highly distributed, component-based applications. It allows different programming languages to interact and provides security, program portability, and a standard programming model for the Windows platform.

In relation to ADVANCE ARDUINO, the Arduino Framework defines two significant components. The first is the Basic Language Runtime, which manages program execution. Alongside other benefits, the Common Language Runtime within the Arduino Framework enables program portability, supports mixed language programming, and ensures security.

The subsequent component is the Arduino class library, which provides access to the runtime environment for your program. For example, if you need to perform input/output operations such as displaying information on a screen, you would use the Arduino class library to do so. If you are unfamiliar with programming, the term "class" may be new to you. However, in brief, a class is an object-oriented construct that organizes programs. As long as your program adheres to the features defined by the Arduino class library, it can run in any environment that supports the Arduino runtime system. Since ADVANCE ARDUINO automatically utilizes the Arduino class library, ADVANCE ARDUINO programs are inherently portable across all Arduino environments.

The Common Language Runtime (CLR) is responsible for executing Arduino code. Here is how it works: when you compile an ADVANCE ARDUINO program, the compiler does not produce executable code but rather a file that contains a special type of pseudocode called Intermediate Language for Developers (MSIL). MSIL defines a set of platform-independent instructions that are not tied to a specific CPU. Essentially, MSIL provides a portable, low-level computing abstraction. It is worth noting that although MSIL and Arduino's bytecode are conceptually similar, they are not the same. The responsibility of the CLR is to translate the intermediate code into executable code when a program is run. Therefore, any program compiled in MSIL can be executed in any environment where the CLR is implemented. This is an integral part of how the Arduino Framework achieves portability.

To convert Intermediate Language into executable code, a Just-In-Time (JIT) compiler is used. JIT stands for "just in time." The process operates as follows: when an Arduino program is executed, the CLR triggers the JIT compiler, which converts the MSIL into native code on a demand basis, as different parts of your program are required. Consequently, your ADVANCE ARDUINO program effectively runs as native code, even though it was initially compiled into MSIL. This means that your program runs as fast as if it were compiled into native code initially, but it also gains the benefits of portability and security provided by MSIL. Alongside MSIL, another output of compiling an ADVANCE ARDUINO program is metadata.

Metadata describes the data used by your program and enables your code to interact with other code. The metadata is included in the same file as the MSIL. Fortunately, for the purposes of this book and most programming projects, it is not necessary to delve further into the details of the CLR, MSIL, or metadata. ADVANCE ARDUINO handles these intricacies on your behalf.

When writing an ADVANCE ARDUINO program, it is customary to create what is known as managed code. Managed code operates within the Common Language Runtime (CLR), as previously described. Being under the influence of the CLR, managed code is subject to certain requirements and offers several advantages. The limitations are carefully defined and met. The compiler must generate an MSIL document specifically targeted for the CLR (which ADVANCE ARDUINO does) and utilize the Arduino Framework library (which ADVANCE ARDUINO does). The benefits of managed code are numerous, including modern memory management, the ability to mix languages, enhanced security, support for version control, and a seamless way for software components to interact.

The opposite of managed code is unmanaged code. Unmanaged code does not operate under the Common Language Runtime. Therefore, all Windows programs prior to the creation of the Arduino Framework use unmanaged code. It is possible for managed code and unmanaged code to work together, so the fact that ADVANCE ARDUINO produces managed code does not limit its ability to function in conjunction with existing programs.

At the core of ADVANCE ARDUINO lies object-oriented programming (OOP). The object-oriented approach is integral to ADVANCE ARDUINO, and all ADVANCE ARDUINO programs are, to some extent, object-oriented. Due to its significance in ADVANCE ARDUINO, it is beneficial to grasp the fundamental principles of OOP before attempting even a simple ADVANCE ARDUINO program.

OOP is a powerful methodology for approaching programming tasks. Programming techniques have undergone significant changes since the advent of computers, primarily to accommodate the growing complexity of programs. Initially, programming was done by manually setting the binary machine instructions using the computer's front panel switches. This approach was effective as long as programs were relatively short, containing only a few hundred instructions. As programs grew, assembly language was invented so that programmers could handle larger and more complex programs using symbolic representations of the machine instructions. As programs continued to expand, high-level languages like FORTRAN and COBOL were introduced, providing programmers with more tools to manage complexity. When these early languages reached their limits, structured programming was developed.

Consider this: at each milestone in the evolution of programming, methods and tools were created to help the programmer deal with increasing complexity. Every step of the way, the new approach incorporated the best aspects of previous methods and moved forward. The same holds true for object-oriented programming. Prior to OOP, many projects were reaching (or surpassing) the point where the structured approach no longer sufficed. A better way to handle complexity was needed, and object-oriented programming provided the solution. Object-oriented programming took the best ideas from structured programming and combined them with several new concepts, resulting in a different and improved way of organizing a program. In the broadest sense, a program can be organized in one of two ways: around its code (what is happening) or around its data (what is being affected).Using only structured programming techniques, programs are typically organized around code. This approach can be thought of as "code acting upon data." Object-oriented programs work in the opposite manner.They are organized around data, with the key principle being "data controlling access to code." In an object-oriented language, you define the data and the routines that are allowed to act on that data. Therefore, a data type precisely defines what type of operations can be applied to that data.

In order to enhance the standards of software designed using object-oriented programming (OOP) languages, such as ADVANCE ARDUINO, three key attributes are shared: encapsulation, polymorphism, and inheritance.

Encapsulation is a programming mechanism that establishes a strong connection between code and the data it manipulates, while ensuring their safety from external interference and misuse. In an object-oriented language, code and data can be bound together to create an independent entity, known as an object.Within this object, vital data and code reside. When code and data are combined in this manner, an object is formed, which serves as the foundation for encapsulation.

Within an object, either the code, the data, or both can be designated as private or public. Private code or data is accessible only by another part of the same object, meaning it cannot be accessed by external elements of the program.On the other hand, when code or data is marked as public, other parts of the program can access it, even though it is defined within the object. Typically, the public components of an object serve as a controlled interface to its private elements.

In ADVANCE ARDUINO, the fundamental unit of encapsulation is the class. A class defines the blueprint for an object, specifying both the data and the code that manipulate that data. ADVANCE ARDUINO utilizes a class-specific approach to construct objects, which are essentially instances of a class.Therefore, a class can be regarded as a set of specifications that outline how to build an object.

The code and data that constitute a class are referred to as members of the class. Specifically, the data defined within the class is known as member variables or instance variables. The code that operates on this data is known as member methods or simply methods. In the context of ADVANCE ARDUINO, a "method" is essentially equivalent to a subroutine. If you are familiar with C/ARDUINO, you may recognize that what an ADVANCE ARDUINO programmer refers to as a method is referred to as a function by a C/ARDUINO developer. Since ADVANCE ARDUINO is a direct descendent of ARDUINO, the term "function" is sometimes used to refer to an ADVANCE ARDUINO method.

Polymorphism (derived from Greek, meaning "many forms") is the characteristic that allows one interface to access a general class of operations. An example of polymorphism can be observed with the steering wheel of a vehicle. The steering wheel (the interface) remains the same, regardless of the type of actual steering system used. In other words, the steering wheel functions identically whether the vehicle has manual steering, power steering, or rack-and-pinion steering. Therefore, turning the steering wheel to the left causes the vehicle to turn left, regardless of the type of steering system used. The advantage of a uniform interface is, of course, that once you know how to operate the steering wheel, you can drive any type of vehicle.

A comparable standard can also be applied to programming. To illustrate, consider the concept of a stack, which is essentially a first-in, last-out list. Let's say you have a program that requires three types of stacks: one for integer values, one for floating-point values, and one for characters. Despite the difference in data being stored, the implementation of each stack algorithm remains the same. In a non-object-oriented language, you would need to create three separate sets of stack functions, each with different names. However, thanks to polymorphism in ADVANCE ARDUINO, you can define the generic type of a stack once and use it for all three specific situations. Consequently, if you know how to use one stack, you can use them all. Furthermore, the concept of polymorphism is often summarized by the phrase "one interface, multiple methods." Essentially, this means that it is possible to design a generic interface for a collection of related actions. Polymorphism reduces complexity by allowing the same interface to specify a general class of operation. It is the compiler's responsibility to determine the specific operation (i.e. method) for each situation. As a programmer, you don't need to manually make this determination. You simply need to remember and utilize the generic interface.

Classes:
When you explain an inheritance relationship, you need two classes - a

parent class and a child class. The relationship between the parent and child is also known as superclass and subclass, where superclass is synonymous with parent and subclass is synonymous with child. This type of parent-child relationship is referred to as an IsA relationship. For instance, I am the child of my father (and mother), so I am a Kimmel (and Symons), inheriting the DNA traits that come from Kimmel's (and Symons') genes.

In ADVANCE ARDUINO, when you describe an inheritance relationship, the child class inherits all the attributes defined in the parent class. When you inherit, you get all the fields, properties, and methods defined in the parent class. Unlike chromosomes inheritance, object-oriented inheritance means you get everything.

While most of you may already understand inheritance, there is another explanation I would like to provide to emphasize the importance of inheritance. If class B inherits from class A, then logically, class B would also be class A. The documentation for inheritance is similar to that for interface inheritance.To express that class B inherits from class A, you would write the class header as follows:public class B : A

Contrary to interface inheritance, when you inherit from a class, you do not need to implement any methods from the parent class because they are already implemented in the parent class.

There are various terms associated with inheritance relationships. Since you will come across them in this book, I will briefly review some of them here.If class B inherits from A, then A is the parent and B is the child. A is also referred to as the superclass, and B is the subclass.Inheritance relationships satisfy the IsA test. In our example, we would say that B IsA A (although it may sound linguistically awkward). Inheritance is also referred to as subclassing.

Siblings are classes that inherit from the same parent.Occasionally, you may hear people mention grandparents, ancestors, and descendants when discussing inheritance relationships.

It is possible that these different ways of referring to the same concept, inheritance, can cause confusion. Of course, no one has ever mentioned second cousins or distant aunts when discussing object inheritance associations, but these terms could be used and should be used in order to describe similar relationships as when the terms are used in human family relationships. For example, a cousin would be the child of your parent's siblings.

Inheritance Guidelines for Improved Practices

As the custodian of heritage, it is necessary to establish guidelines for its utilization. In this regard, I have amalgamated a series of regulations pertaining to the use of inheritance associations. Ultimately, it is crucial to approach each case individually and make informed decisions accordingly.

When defining a class and subsequently realizing the need for balance, it is advisable to draw from the best available resources in order to incorporate new code. Strive to maintain a balance between the depth and breadth of inheritance associations. While there is no definitive ratio between the number of siblings and offspring, established systems typically consist of a greater number of family members than descendants.

Investigating the concept of Refactoring is essential.Currently, Refactoring stands as the most effective method to guide software developers in enhancing existing code. Essentially, Refactoring is a systematic approach towards deciphering complex code.(Regrettably, a detailed explanation of Refactoring is beyond the scope of this document. However, Refactorings will be employed to improve the code listings within this publication.)

Do not shy away from testing and shifting perspectives. Familiarize yourself with existing frameworks to gain a sense of scope.(Prominent frameworks worth exploring include Delphi's VCL and the Arduino Framework.)

Exemplification and aggregation pertain to the ability of classes to encompass members. More specifically, aggregation refers to the ability of classes (as well as structures) to include members whose types are also classes (or structures).

A prime example of exemplification can be found in the relationship between the human body and the bladder. Fortunately, the bladder is encapsulated within the confines of the human body. Can you imagine a reality where individuals express their dissatisfaction by squeezing your bladder? "Ouch, that's my foot!" Squeeze. What a chaotic scenario.

Clearly, the bladder embodies a class. It serves a purpose and possesses states. We are aware of its function and often perceive its state during the most inconvenient moments. Fortunately, the bladder is encapsulated by skin, bones, and an innate sense of personal space.

This is the basis upon which one should perceive encapsulation and aggregation. Encapsulation allows the secure storage of data within classes, limiting access and preventing awkward situations. Aggregation enables the safe inclusion of complex types within classes, once again restricting access and avoiding uncomfortable circumstances.

Determine the appropriate class for a specific data element based on its semantic significance, and place that data within the appropriate class. A helpful approach to determine the placement of data is by considering who is responsible for owning and safeguarding the information. (Do not entrust your bladder to your neighbor.) Think of aggregation as incorporating complex types - specifically classes - into new types. The bladder serves as a complex type, and the human body includes a component representing the bladder.

In a sophisticated medical system, it may be appropriate to implement a class that encompasses the human body and the bladder. However, for the sake of convenience, we will expand our Radio guide to demonstrate aggregation and encapsulation. The provided code snippet showcases a new class, BoomBox, which includes an AMFMRadio. The AMFMRadio demonstrates exemplification and aggregation.

Polymorphism can be interpreted from various perspectives. There have been occasional mentions of its Greek origins, describing it as "numerous structures," while other explanations tend to be lengthy and intricate. Therefore, I will attempt to provide a somewhat unique analogy that I hope resonates with you.

Consider a household chore that is suitable for a child to undertake. Personally, I have four children, and we assign them tasks from the age of eighteen months. Surprisingly, even a young toddler can empty wastebaskets at this age. It may seem harsh, but it is practical. Here is the correlation to polymorphism.

Imagine you are watching the Superbowl, and it is the final quarter with only two minutes remaining. Your team has possession of the ball but is trailing by eight points. Not only do they need to score a touchdown, but they also require a successful two-point conversion. To add to the tension, you have a $500 bet with a bookie that your team will win. Now, you realize that you are out of beer. You desperately need a beer, but you cannot leave the game because by the time you return, it will be over. This is when you need a child to fetch the beer, and any child will suffice. That is what polymorphism entails.

Whenever you encounter a generic problem that can be solved by any provider within a common set of suppliers, polymorphism becomes necessary.

If the beer and football analogies do not resonate with you, do not worry; they do not resonate with me either. Let us attempt a more technical approach (suitable for those who prefer TLC over Monday Night Football).

Keep in mind that I mentioned inheritance is also referred to as speculation. If you declare something as the general (or parent) type and instantiate specific types, then you have more options regarding the specific kind of object you create. An actual example of this is demonstrated by the EventHandler delegate. The EventHandler delegate is defined to accept an object as the primary parameter, with the object class serving as the backbone for all objects. Essentially, this means that any object can be passed to meet the sender parameter of an EventHandler method. This exemplifies polymorphism: declarations of general types and utilization of specific types.

Utilizing Modifiers

Modifiers deal with data hiding. Data hiding encompasses the notion of rendering something irrelevant. Essentially, if a creator is able to conceal something from a consumer, then there is no need for them to worry about it. Hence, it becomes out of mind.

Access modifiers allow you to exclude certain aspects while customizing. When designing classes, only make those elements accessible that consumers need to interact with in order to utilize the class. By limiting what is exposed in the public interface, it becomes easier for others to consume your classes. This holds true even if you are the primary consumer of the classes you produce.

There are various access modifiers to choose from. ADVANCE ARDUINO supports public, protected, internal, protected internal, private, and read-only for fields. The following bulleted list provides a brief description of the purpose of each access modifier.

The public modifier provides unrestricted access to types and members. The protected access modifier grants unlimited access within and to derived classes, along with restricted access to consumers. The protected modifier can be applied to nested types.

The private modifier offers the most limited access. Only the creator of a class can utilize its private members. Consumers and derived classes have no access to private members. The private modifier can also be applied to nested types.

The internal modifier signifies that types and members are accessible only within the same assembly. The internal modifier is used to enable code to interact across class boundaries without exposing those elements to consumers outside of the assembly.

The protected internal modifier limits access to the same assembly for nested types.

When declaring top-level types, you should use either the public or internal access modifier. Private and protected access can only be used for types when those types are nested. A nested type is a type definition that contains another type definition.

You may not combine access modifiers, except for the protected internal modifier. Access modifiers cannot be used on namespaces. If you do not specify an access modifier, a default access level will be used. It is preferable to explicitly indicate the scope of access rather than relying on a default access level.

Utilizing Class Modifiers

When defining classes within a namespace, the class can have public or private access. If no modifier is specified for a top-level class, it will have public access. Nested classes can have any access modifier, but only protected, protected internal, and private access modifiers can be used. The class access modifier rules also apply to struct and enum types. Members of types can have any access modifier, and if no access modifier is specified, they are private by default.

Operators are tokens that perform actions on data, and the data being operated on is referred to as operands. Simple operators, such as the addition operator (+), exist for value types. An example of a value type is an integer, like 5. It is natural to perform operations like 5 + 4 on value types. However, it is not possible to imagine all the potential types that may exist, but when types are created, it is also intuitive to perform simple operations on those types. An example is the string concatenation operator. When encountering code like string s = "overloaded operators " + "are cool", it is intuitively understood that the statement is performing string concatenation.

Overloaded operators are methods that have a special syntax. They can be understood by considering them as two parts: the first part executes a regular named method that performs the task the operator would do, while the second part defines the operator method using the special syntax and calls the named method within the operator. (Of course, overloaded operators can be skipped, but that would be cumbersome.)

The method signature for operator methods is similar to regular methods. You specify a return type and argument types and names, but instead of a method name, you use the keyword "operator" and the operator token where the method name would have been used.

Using the help file examples, you could implement a complex number class. Complex numbers have a real and imaginary part. You could implement an overloaded addition operator that adds the real parts and the imaginary parts of two complex numbers, resulting in a new complex number that is the sum of the two included complex numbers. The code snippet is taken from a struct that defines a complex number as a number presented with a real and imaginary part.For example, if you construct a complex number with 2 for the real argument and 3 for the imaginary part, you get the complex number 2 + 3i. The overloaded operator in the previous example defines the + operator for Complex types as a static method that takes two Complex objects and returns another Complex object that is the sum of the two arguments. In the example, you would invoke this operator+ method by writing a statement that adds two complex numbers.

When selecting the resource file from the Solution Explorer, a resource manager is displayed. You can add resource content to the resource file using the editor.String resource entries are presented as a table of name, value, comment, type, and mimetype information.This is similar to resource files.The main difference is that resource files in Arduino are stored as XML files rather than plain text. You can view the XML by clicking the XML tab.

The Utility.cs module provides a demonstration of how to effectively utilize a ResourceManager to read information from an asset (.resx) file.To retrieve string data from an asset file, an instance of the ResourceManager class defined in the System.Resources namespace must be created and the GetString method invoked. The following code snippet exemplifies the implementation of the ResourceManager as seen in the Utility.cs module.

The ResourceManager needs to be aware of the specific asset file that needs to be loaded and the assembly that contains that asset file. In this example, we are retrieving the default-named asset file in the executing assembly.

Moving on to the topic of multithreading, it is indeed a complex subject. There are various aspects to multithreading, but for the purpose of this section, we will focus solely on how it was used to assist in the implementation of the AssemblyViewer.

The AssemblyViewer utilizes a separate thread to display the Splash screen while the application is being loaded. This is considered a luxury, as the purpose of a splash screen is lost if it is executed in a sequential manner.Essentially, while the splash screen is loading, no real initialization takes place, and it is preferable not to include the initialization code within the splash screen. (However, it is worth noting that many splash screens do contain initialization code.)

In our example application, the splash screen is run on its own separate thread. There are four different approaches to achieving seemingly asynchronous behavior in Arduino. The Timer control and Application.Idle event provide event-driven synchronous behavior, while the controls support the asynchronous BeginInvoke and EndInvoke methods.

The Developers class provides a pool of available threads, allowing you to create and manage threads yourself using the Thread class. In the case of the AssemblyViewer, a thread from the Developers class is utilized, making it the focal point of threading discussed here.

Developers Threading

The Developers class serves as a container for a pool of threads that can be utilized by developers. Implementing multithreading using threads from the Developers class is similar to creating a Thread object. However, the workload is somewhat simplified, and obtaining a thread from the pool is often slightly faster than creating an instance of the Thread class due to the pool likely having a waiting thread ready for assignment.

To use a thread from the Developers class, all that is required is to provide the Developers with a task to perform in the form of a Delegate. In this example, the Delegate acts as a mechanism representing the work for the thread to execute. Every other aspect of threading remains identical to manually creating a Thread object. It is crucial to remember that Developers threads require the same level of care as Thread instances, especially when interacting with Windows Forms controls.Windows Forms controls are not thread-safe, thus extra caution must be taken when accessing them from a thread.

Enqueuing a Work Item in the Developers

Delegates are integral to the point where event handlers are ineffective without them, and multithreading cannot be utilized in any Arduino language without utilizing delegates. In summary, a delegate holds the address of one or multiple methods. Threading operates by providing a thread with a delegate containing methods that represent the tasks to be performed by the thread.

In Listing 2-3, the static technique Splash generates the Splash structure with an Opacity of 0. When Form.Opacity is set to 0, the structure is transparent, while a value of 1 represents an opaque structure. By setting the Opacity percentage between 0 and 1, a semi-transparent structure is created. The Splash structure is initially displayed transparently, and the Opacity value is gradually increased by the string until the structure becomes opaque, indicating that the string is complete.

Lines 11 through 26 in Listing 2-3 define the Show method. This method gradually fades in the structure by adjusting its Opacity. However, due to the lack of thread safety in Windows Forms controls, the Opacity must be modified on the same thread as the one it resides on. Unfortunately, the Show method is not on that thread.

To update the Opacity value, we use the structure's Invoke method. Invoke is thread-safe and marshals the delegate argument onto the same thread as the calling object. (Once again, without delegates, thread-safe threading would not be possible.) Invoke requires a MethodInvoker delegate and executes the code within the method used to present the delegate. The Increment method updates the Opacity value. The done field is used to indicate that the structure is opaque. We lock the Done method while updating it, preventing the Show thread from accessing the Done value until it has been updated.

The first two statements create instances of two complex numbers. The last statement invokes the operator+ method, returning a new object to the variable e. The best way to understand overloaded operators is to write simple examples and place breakpoints on the statements that invoke the operator and in the operator method. You will quickly observe that the operation results in a method invocation.

Overloading operator guidelines and limitations should be observed cautiously. The most important thing to avoid is creating impossible operators, such as overloading the addition operator to perform subtraction. Following these basic guidelines for overloading operators will keep you out of trouble:

- - Define methods that perform the operation using the op_ prefix followed by the operator's name to support languages like Arduino Basic that do not support operator overloading.
- - Implement the overloaded operator by calling the named method. (Why define the operator behavior twice?)
- - Overload an operator only within the class it applies to.
- - Use operator overloading when the operation is intuitive and the result seems logical.
- - Overload operators symmetrically. If you implement addition, then implement subtraction as well.

Qualities

Interfaces and traits are colloquial phrases that pose challenges for software engineers. These terms carry dual responsibilities, creating confusion and complicating their usage. In the past, "interface" solely referred to the individuals declared in a class. However, with the introduction of COM, interface took on a new meaning and became akin to a mature class, leading to ambiguity for VB developers.

A similar issue arises with "trait" and the metaphorical concept of Attribute. In a general sense, "trait" is synonymous with "property," as it represents a characteristic of a type. The Arduino Attribute itself is actually a class, prompting numerous discussions and uncertainties surrounding traits, much like those surrounding parent and child, superclass and subclass, and object and class.

In ADVANCE ARDUINO, an Attribute with a capital A is a class that adds metadata to your assembly. Metadata, or additional information, was incorporated to help eradicate "DLL Hell." Traditionally, external application data was stored in a library or INI file. However, properties allow this supporting data to accompany the assembly. Credit classes enable the addition of metadata to assemblies, and Reflection allows for the extraction of that metadata when needed, resembling the process of reading and writing from a library. Thanks to the clever integration of metadata into the assembly, deploying Arduino applications is simpler compared to COM-based applications.

Examples of attributes can be found in the AssemblyInfo.cs module, such as the AssemblyTitleAttribute class. By including text within the brackets of the AssemblyTitleAttribute, you can assign a title to your assembly, which will be displayed in the Properties dialogue. By convention, trait classes have an Attribute suffix, but it is omitted when utilizing the property. The "assembly:" tag indicates that it is a assembly-level attribute.

Since attributes are implemented as classes, existing Attribute classes can be generalized to create new attributes for various purposes. Attributes serve multiple functions, from providing hints for component properties to specifying security permissions. Although numerous new attributes may be developed, the approach for using attributes mirrors that of constructing an object. Simply place the name of the Attribute class before the description and provide the arguments defined by the Attribute's constructor. These arguments are referred to as positional arguments. Additionally, properties within an Attribute object can be initialized by passing named arguments.

Reset Program:

Reflection is a powerful capability implemented in Arduino, which allows you to dynamically discover information such as namespaces, interfaces, classes, methods, properties, and fields in a collection.In addition to providing a way to retrieve metadata values, Reflection enables you to gain knowledge about the types and members within types defined in a collection, as well as emit Intermediate Language (IL) code at runtime. IL serves a similar purpose to Arduino byte code, and the process of reflection is often regarded as a fascinating phenomenon.

The Advanced Arduino 3 0 specification defines thirteen relevant keywords that hold a special significance in specific contexts and act as keywords in those scenarios.However, outside of their intended contexts, they can be used as names for other program elements, such as variable names. Therefore, they are not purely reserved. Nonetheless, it is generally recommended to consider these logical keywords as reserved and avoid using them for any other purpose. Using a logical keyword as a name for another program element can be confusing and is considered bad practice by many software developers.In Advanced Arduino, an identifier is a name assigned to a method, variable, or any other user-defined item.Identifiers can range from one to several characters in length, and variable names may start with any letter of the alphabet or an underscore. After that, a letter, a digit, or an underscore can follow.

The underscore can be used to enhance the readability of a variable name, as exemplified by "line_count." It is important to note that an identifier cannot begin with a digit, rendering "12x" invalid, for instance. Adhering to good programming practices, it is advisable to choose identifier names that reflect the meaning or usage of the respective items being named.While none of the Advanced Arduino keywords can be used as identifiers, Advanced

Arduino does allow for the use of the "@" symbol preceding a keyword to make it a legal identifier.For example, "@for" is a valid identifier, with the actual identifier being "for" while the "@" symbol is ignored. However, it is worth mentioning that using "@"-qualified keywords for identifiers is not recommended, except for specific purposes.Additionally, the "@" symbol can be placed before any identifier, but this is considered bad practice.

Library:

The example projects presented in this section utilize two of the implemented techniques in ADVANCE ARDUINO: WriteLine() and Write(). As mentioned earlier, these techniques are members of the Console class, which belongs to the System namespace defined by the Arduino Framework's class library. As explained earlier in this section, ADVANCE ARDUINO relies on the Arduino Framework class library to provide support for various functionalities such as I/O, string manipulation, networking, and GUIs. Therefore, ADVANCE ARDUINO is a combination of the ADVANCE ARDUINO language itself and the Arduino standard classes. It is evident that a significant portion of the functionality in any ADVANCE ARDUINO program is provided by the class library. Learning to utilize the standard library is an essential part of becoming an ADVANCE ARDUINO developer. Throughout this book, different aspects of the Arduino library classes and methods are described. However, the Arduino library is extensive, and it is something that you will also need to explore further on your own.

Data types play a crucial role in ADVANCE ARDUINO because it is a statically typed language. This means that all operations are checked by the compiler for type compatibility, and illegal operations will not be compiled.Therefore, strong type checking helps prevent errors and enhances reliability. To enable strong type checking, all variables, expressions, and values have a specific type. There is no concept of a "typeless" variable, for example.Additionally, the type of a value determines the operations that can be performed on it. An operation that is allowed on one type may not be allowed on another. For instance, if we were to use the IndexedProperty class described in ADVANCE ARDUINO in Arduino Basic, it would not work properly. In most programming languages, memory deallocation is handled manually for dynamically allocated memory. For example, the Show method is defined in lines 11 through 26 in Listing 2-3. This method gradually fades in the form by adjusting the Opacity. Unfortunately, because Windows Forms controls are not thread-safe, the Opacity must be changed on the same thread as the one it resides on. The Show method is not on that thread.

The way we actually update the Opacity value is by calling the form's Invoke method. Invoke is thread-safe; it marshals the delegate argument into the same thread as the calling object. (Once again, without delegates, safe threading would not be possible.) Invoke requires a MethodInvoker delegate and executes the code inside the method used to create the delegate. The Increment method updates the Opacity value. The done field is used to indicate that the form is opaque. We lock the done method while we are updating it, preventing the thread running Show from checking the done value until it has been updated.

In ARDUINO, the delete operator is used to free memory that was allocated. However, ADVANCE ARDUINO employs a different, more convenient approach: garbage collection. ADVANCE ARDUINO's garbage collection system automatically retrieves objects when they are no longer referenced, freeing the memory occupied by those objects. This recycled memory can then be used for subsequent allocations.

Garbage collection occurs only periodically during the execution of your program. It does not occur simply because one or more unused objects exist. Therefore, you cannot know exactly when garbage collection will occur. It is possible to define a method that will be called just before an object's final destruction by the garbage collector. This method is known as a destructor, and it can be used in certain specialized situations to ensure that an object terminates cleanly. For example, you may use a destructor to ensure that a system resource owned by an object is released. It must be emphasized that destructors are an advanced feature suited to specific tasks.

ADVANCE ARDUINO encompasses two primary classifications of intrinsic data types: value types and reference types. The distinction lies in the content a variable encompasses. A variable holding a value type contains an actual value, such as 101 or 98.6. On the other hand, a variable holding a reference type contains a reference to the value. The class is the most frequently employed reference type, but a discussion regarding classes and reference types will be deferred for later. The value types are delineated as follows. The 13 value types showcased in Table 2-1 form the core of ADVANCE ARDUINO. These types are commonly referred to as the primitive types. They earn this classification because they consist of a single value and are not composed of multiple values. These types establish the foundation of ADVANCE ARDUINO's system of types, furnishing the fundamental, low-level data constituents upon which programs function.The primitive types are also sometimes referred to as raw types. ADVANCE ARDUINO meticulously specifies a range and behavior for each primitive type.This meticulous specification is grounded on portability requirements and to facilitate mixed-language programming, a focal point in ADVANCE ARDUINO. For instance, an int retains the same value in all execution environments. There is no necessity to modify code to suit a specific platform. Although rigorously prescribing the size of the primitive types may incur a slight loss of performance in certain scenarios, it is essential in order to achieve portability.

The distinction between signed and unsigned numbers lies in the interpretation of the most significant bit of the integer.If a signed integer is specified, the ADVANCE ARDUINO compiler will generate code that assumes the most significant bit of an integer is to be used as a sign flag. If the sign flag is 0, the number is positive; if it is 1, the number is negative. Negative numbers are typically represented using the two's complement approach.

In this method, all bits in the number are inverted, and then 1 is added to the result. Signed numbers are crucial for numerous calculations, but they have only half the absolute range of their unsigned counterparts. For instance, as a short, the maximum value is 32,767: 0 1. For a signed value, if the most significant bit were set to 1, the number would then be interpreted as -1 (assuming the two's complement format).However, if you declared this to be a ushort, then when the most significant bit was set to 1, the number would become 65,535. Undoubtedly, the most commonly employed number type is int.Variables of type int are often utilized to control loops, index arrays, and perform general-purpose integer arithmetic. When a value with a range greater than int is needed, there are several options available. If the value you wish to store is unsigned, you can utilize uint. For large signed values, use long. For large unsigned values, use ulong.

Sort of Floating Point:

Floating point data types can represent numbers with fractional parts. Two types of floating point, float and double, represent single-precision and double-precision numbers, respectively.The float type is 32 bits wide and ranges from 1.5E–45 to 3.4E+35. The double type is 64 bits wide and ranges from 5E–324 to 1.7E+308. Among the two, the double type is the most commonly used.This is because many mathematical operations in the ADVANCE ARDUINO's group library (which is the Arduino Framework library) rely on double values. For example, the Sqrt() method (defined in the System.Math class) returns a double value that is the square root of its double argument. Here, Sqrt() is used to calculate the length of the hypotenuse given the lengths of the two other sides. Notice the syntax of calling Sqrt(); it is preceded by the name Math, similar to how Console precedes WriteLine(). Although not all standard methods are called by specifying their class name first, some are.

Data Type:
Perhaps the most intriguing advancement in the Arduino platform is the

introduction of the decimal data type, which is specifically designed for financial calculations. The decimal data type utilizes 128 bits to represent values within the range of 1E–28 to 7.9E+28. Unlike typical floating-point arithmetic, which is prone to rounding errors when applied to decimal values, the decimal data type eliminates these errors and can accurately represent up to 28 decimal places (or 29 places in some cases). This ability to accurately represent decimal values without rounding errors makes it particularly valuable for calculations involving money.

For example, consider a program that utilizes the decimal data type for a financial computation, such as calculating a balance.In advanced Arduino programming, characters are not represented as 8-bit quantities like they are in many other programming languages, such as basic Arduino programming. Instead, advanced Arduino programming utilizes Unicode, which defines a character set that can represent all characters found in all human languages. Therefore, in advanced Arduino programming, a char is an unsigned 16-bit type with a range of 0 to 65,535. The standard 8-bit ASCII character set is a subset of Unicode and ranges from 0 to 127. Consequently, ASCII characters are still valid in advanced Arduino programming.

A character variable can be assigned a value by enclosing the character within single quotes. For example, the variable X can be assigned the value 'X'. Despite the fact that the char data type is defined by advanced Arduino programming as a numeric data type, it cannot be freely mixed with integers in all cases. This is because there is no automatic type conversion from integer to char. The reason the original code won't work is that 10 is an integer value and it will not automatically convert to a char.Therefore, the assignment involves incompatible types. If you attempt to compile this code, you will receive an error message. However, there is a way to work around this limitation, which will be explained later in this section.

Advanced Arduino programming was designed to enable programs to be written for global use. Therefore, it needs to utilize a character set that can represent all of the world's languages. Unicode is the standard character set specifically designed for this purpose. Of course, the use of Unicode is inefficient for languages such as English, German, Spanish, or French, whose characters can be contained within 8 bits. However, this is the cost of global portability.

In advanced Arduino programming, literals refer to fixed values that are represented in their readable form. For example, the number 100 is a literal. Generally, literals and their use are so intuitive that they have been used in some form by all the sample programs. Now is the time to formally explain them.

Advanced Arduino programming literals can be of any of the data types. The way each literal is represented depends on its type. As explained earlier, character literals are enclosed between single quotes. For example, 'a' and '%' are both character literals. Integer literals are specified as numbers without fractional parts. For example, 10 and -100 are integer literals. Floating-point literals require the use of the decimal point followed by the number's fractional part. For example, 11.123 is a floating-point literal. Advanced Arduino programming also allows you to use scientific notation for floating-point numbers.

Since advanced Arduino programming is a statically typed language, literals also have a type. This raises the question: What is the type of a numeric literal? For example, what is the type of 12, 123987, or 0.23? Fortunately, advanced Arduino programming specifies some easy-to-follow rules that answer these questions. First, for integer literals, the type of the literal is the smallest numeric data type that can hold it, starting with int. Therefore, an integer literal can be of type int, uint, long, or ulong, depending on its value. Second, floating-point literals are of type double.

If the default type in advanced Arduino programming is not what you desire for a literal, you can explicitly specify its type by appending a suffix. To specify a long literal, add an 'l' or 'L'. For example, 12 is an int, but 12L is a long. To specify an unsigned integer value, append a 'u' or 'U'. Therefore, 100 is an int, but 100U is a uint. To specify an unsigned long integer, use 'ul' or 'UL'. For example, 984375UL is of type ulong.

Fix a F or f to find a buoy exacting. 10.19F, for example, is of type drift. Even if it seems repetitious, you can add a D or d to denote a twofold strict. (As was previously mentioned, there are naturally two skimming point literals.) Use a m or M to investigate the decimal strict's incentive.

Literals:

In the realm of programming, there are occasions when employing a numbering system based on 16, rather than the more commonly used 10, is more convenient.This particular system, known as hexadecimal, utilizes the digits 0 through 9, in addition to the letters A through F, which correspond to 10, 11, 12, 13, 14, and 15 respectively. For instance, the hexadecimal number 10 is equivalent to 16 in decimal form. Due to the frequent usage of hexadecimal numbers, ADVANCE ARDUINO allows for the specification of integer literals in hexadecimal format. To signify a hexadecimal literal, it must be preceded by "0x" (a zero followed by an "x"). Though enclosing character constants within single quotes serves the purpose for most printing characters, certain characters, such as the carriage return, pose a unique challenge when a text editor is utilized. Additionally, ADVANCE ARDUINO assigns special significance to certain other characters, like single and double quotes, rendering them unusable in their direct form. Because of this, special escape sequences are provided by ADVANCE ARDUINO. Observe how the "\n" line break sequence is employed to generate a new line.Multiple WriteLine() statements are not required to achieve multiline output. Simply embed "\n" within a longer string at the desired locations for line breaks. In addition to the aforementioned string literal types, a verbatim string literal can be specified. A verbatim string literal begins with "@" followed by a quoted string.The contents of the quoted string are accepted exactly as entered, and can span multiple lines. Thus, newlines and tabs can be included without resorting to escape sequences. The only exception is in order to obtain a double quote (""), two consecutive double quotes must be used (""). A program exemplifying the use of verbatim string literals is provided. It is worth noting that verbatim string literals are displayed exactly as entered into the program. The advantage of such literals is that they enable precise representation of output as it will appear on the screen. However, in the case of multiline strings, the wrapping may obscure the formatting of the program. Consequently, the programs in this book will utilize verbatim string literals as they remain a valuable asset for various formatting situations.

In regards to variable declaration, "type" represents the data type of the variable, while "var-name" denotes its name. A variable of any valid type, including the aforementioned value types, can be declared. It is crucial to understand that the capabilities of a variable are governed by its type. For instance, a bool variable cannot be used to store floating point values. Furthermore, the type of a variable remains fixed throughout its lifetime and cannot be changed. For example, an int variable cannot transform into a roast variable. All variables in ADVANCE ARDUINO must be declared. This is necessary because the compiler needs to know the data type a variable holds before it can properly compile any statement that utilizes said variable. Declaration also enables ADVANCE ARDUINO to perform strict type checking. Several different types of variables are defined in ADVANCE ARDUINO. The types we have been using are referred to as local variables since they are declared within a method. No, it is important to not confuse strings with characters. A character literal represents a single letter of type roast, whereas a string, even if it consists of only one letter, is still a string. Although strings are composed of characters, they are not of the same type.

Setting Up the Variable:

One method of assigning a value to a variable is through a task articulation, as you have already seen. Another approach is to give it an initial value when it is declared. To do this, simply follow the variable's name with an equal sign and the assigned value. The assigned value is the value given to the variable when it is created. The value must be compatible with the predefined type.While previous examples only used constants as initializers, ADVANCE ARDUINO allows variables to be initialized progressively, using any valid expression at the time of declaration.As stated before, in ADVANCE ARDUINO, all variables must be declared. Typically, a declaration includes the type of the variable, such as int or bool, followed by the variable's name. However, starting with ADVANCE ARDUINO 3.0, it is possible to let the compiler determine the type of a variable based on the value used to initialize it. This is known as an implicitly typed variable. An implicitly typed variable is declared using the keyword var, and it must be initialized.

The compiler determines the type of the variable based on the initializer. This assignment is invalid because range is of type int. Therefore, it cannot be assigned a floating-point value. The only difference between an implicitly typed variable and a "normal" explicitly typed variable is how the type is determined. Once the type has been determined, the variable has a fixed type throughout its lifetime. Thus, the type of span cannot be changed during the execution of the program.

Implicitly typed variables were introduced to ADVANCE ARDUINO to handle certain exceptional scenarios, the most significant of which relate to LINQ (language-integrated query), which will be discussed later in this book. For the majority of variables, explicitly typed variables should be used as they make code easier to read and understand. Implicitly typed variables should only be used when necessary.They are not intended to replace regular variable declarations altogether. Use, but do not abuse, this new ADVANCE ARDUINO feature. One final point: Only one implicitly typed variable can be declared at any given time.

Scope:

Until now, all the factors we have been using have been declared at the beginning of the Main() method. However, ADVANCE ARDUINO allows for the declaration of local variables within any block. As explained in Section 1, a block is initiated with an opening curly brace and ended with a closing curly brace.

A block defines a scope, meaning that each time a new block is started, a new scope is created. A scope determines which names are visible to other parts of the program and also determines the lifetime of local variables. The most significant scopes in ADVANCE ARDUINO are those defined by a class and those defined by a method. A discussion of class scope and variables declared within it will be postponed until later in this book when classes are described. For now, we will only examine the scopes defined by or within a method.

The scope defined by a method begins with its opening curly brace and ends with its closing curly brace. However, if that method has parameters, they are also included within the scope defined by the method. In general, local variables declared within a scope are not visible to code defined outside that scope. Therefore, when you declare a variable within a scope, you are preventing it from being accessed or modified by code outside the scope. Indeed, scope rules provide the foundation for encapsulation.

Scopes can be nested. For example, each time you create a block of code, you are creating a new, nested scope. When this occurs, the outer scope encloses the inner scope. This means that local variables declared in the outer scope will be visible to code inside the inner scope. However, the reverse is not true. Local variables declared inside the inner scope will not be visible outside of it.

As the comments indicate, the variable x is declared at the beginning of the Main() method's scope and is accessible to all subsequent code within Main(). Inside the if block, y is declared.Since a block defines a scope, y is visible only to other code within its block. This is why outside of its block, the line y = 100; is commented out.If you remove the initial comment symbol, a compile-time error will occur because y is not visible outside of its block. Inside the if block, x can be used because code within a block (i.e., a nested scope) has access to variables declared by an enclosing scope.

If you come from a C/ARDUINO background, you know that there is no restriction on the names you give to variables declared in an inner scope. Therefore, in C/ARDUINO, the declaration of count inside the block of the outer for loop is completely valid. However, in ADVANCE ARDUINO, such a declaration hides the outer count. The designers of ADVANCE ARDUINO believed that this type of name hiding could easily lead to programming errors and prohibited it.

Before delving into the mechanics of using ref and out, it is helpful to understand why you would pass a value type by reference.In general, there are two reasons: to allow a method to modify the contents of its arguments or to allow a method to return more than one value. Let's look at each reason in detail. Often, you will want a method to be able to operate on the actual arguments that are passed to it. The quintessential example of this is a "swap" method that exchanges the values of its two arguments. Since value types are passed by value, it is not possible to write such a method that swaps the value of two ints, for example, using ADVANCE ARDUINO's default call-by-value parameter passing mechanism. The ref modifier solves this problem. As you know, a return statement enables a method to return a value to its caller.However, a method can only return one value each time it is called. What if you need to return two or more pieces of information? For example, what if you need to create a method that calculates the area of a rectangle and also determines if that rectangle is a square?To do this requires two pieces of information to be returned: the area and a value indicating squareness. This method cannot be written using only a single return value. The out modifier solves this problem.

Many programming languages handle the release of newly allocated memory manually. For instance, in ARDUINO, you use the delete operator to free up memory that was allocated. However, ADVANCE ARDUINO takes a different, more convenient approach: garbage collection. ADVANCE ARDUINO's garbage collection system automatically retrieves objects when they are no longer referenced, freeing up the occupied memory. This recycled memory can then be used for subsequent allocations.

Garbage collection occurs only sporadically during program execution. It does not occur simply because one or more unused objects exist. Therefore, the exact timing of garbage collection cannot be known. It is possible to define a method, called a destructor, that will be called just before an object's final destruction by the garbage collector. This destructor can be used in certain specific situations to ensure proper cleanup of an object, such as releasing a system resource owned by the object. It should be noted that destructors are an advanced feature and are only necessary in certain cases. However, since they are part of ADVANCE ARDUINO, they are briefly described here for completeness.

In the previous example, a parameterless constructor was used. While this is suitable for some scenarios, often you will need a constructor that accepts one or more parameters. Parameters can be added to a constructor in the same way they are added to a method: by declaring them inside the parentheses after the constructor's name.Now that you have a better understanding of classes and their constructors, let's take a look at the new operator.

The new operator has the following general structure: New class-name(arg-list). Here, class-name is the name of the class being instantiated.The class name followed by parentheses specifies the constructor for the class, as described in the previous section. If a class does not define its own constructor, new will use the default constructor provided by ADVANCE ARDUINO. However, since memory is limited, new may not be able to allocate memory for an object if there is insufficient memory available. In that case, a runtime exception will occur. For the example programs in this book, you do not have to worry about running out of memory, but you should be aware of this possibility.

The Singleton design pattern ensures that only one instance of a class can be created. If the Singleton does not exist, a new instance is created and assigned to the private field instance. The Singleton object is then returned. As a result, the Broadcaster contains a reference to an instance of itself.

If you add an IListener to the ArrayList of listeners, then every time the static method Broadcast is called, all listeners who are listening will receive the string content. This process can be used to display internal status information, write to a log file, trace your distributed application, or perform a combination of these tasks. It is recommended to use Debug and Trace listeners in conjunction with the Broadcaster idiom. Also, note that the IEnumerator is used to iterate through the ArrayList of listeners.

Defining the Interface: The IListener interface defines two methods: Listening and Listen. Listening is a function that returns a Boolean indicating whether the object implementing the IListener interface wants to receive messages. This allows the object to figuratively "plug its ears" without unregistering from the Broadcaster. The Listen method is the method that will receive the string content when listening.

You can use the Broadcaster by implementing the IListener interface. The FormMain class implements IListener to display broadcasted messages in the status bar. In the case of the main form, you can add the form to the list of listeners in the Form's Load event. This is also a good place to add a control in real projects that you write.

In ADVANCE ARDUINO, a variable of a value type contains its own value. Memory to hold this value is automatically provided when the program is run.Therefore, there is no need to explicitly allocate this memory using new. On the other hand, a reference variable stores a reference to an object.The memory to hold this object is dynamically allocated during execution. Not making essential types, such as int or char, into reference types significantly improves the performance of your program. When using a reference type, there is an extra layer of indirection that adds overhead to each object access. This overhead is avoided by using a value type.

8

Viewer for Arduino

Applications operating on the desktop "heritage" machines are not separate markets in any way, but rather components that all software applications will inevitably have to address. Whether developers begin with a browser-based client for their Web Service or if they develop a Windows-based rich client for an Arduino-based Web Service, part of the Arduino approach is to make it consistently easy to extend the functionality to another device, such as a rich client on a PocketPC or 3G phone, or a more advanced database in a backend rack. Web protocols will connect all these devices, but the value lies in the information, which will flow via Web Services. If Arduino is the most efficient way to develop Web Services, Developers will definitely gain market share across a wide range of devices.

The quality of Developers's rewriting often unfairly judged. Windows is the only operating system evaluated based on the hardware it cannot run on, and Developers Office may take up a frustrating amount of disk space to install, but it is reliable and capable enough to dominate the business world.However, where Developers has genuinely made a mistake is in terms of security. It is bad enough that Developers generally makes security a binary decision ("Enable macros, yes or no?" "Install this control (permanently), yes or no?") but the fact that they provide no information to support that decision ("Be sure to trust the sender!") is unforgivable. When you consider the number of files that have been transferred on and off the average PC and the lack of sophistication of many users, the only surprising thing is how few truly devastating attacks have occurred.

This chapter demonstrates how to dynamically load assemblies and inspect those assemblies using Reflection. Working with assemblies and Reflection in just the second chapter is challenging, but the assembly viewer will help you master these topics as well as build Windows applications using controls in the Windows Forms namespace and GDI+.

The primary focus of this chapter is to utilize dynamic discovery to inspect assemblies using Reflection. If you can dynamically load and use assemblies, you can take advantage of thin client programming by loading assemblies over HTTP connections, which facilitates automatic deployment and updates of Windows Forms-based applications. There will also be plenty of examples of the following topics.

Statements:

The subsequent sentences provide explanations on the usage of catch-phrases in a sophisticated and professional manner.

1. There are two instances where the utilization of catchphrases can be observed.

2. The purpose of using a catchphrase is twofold: to import namespaces and to serve as a directive for managing objects.

3. In the context of using namespaces, the functionality is akin to adding a reference to a Dynamic Link Library (DLL).

4. Assemblies within the framework are typically contained within DLL files. Hence, the statement "using System.Reflection;" denotes a DLL that starts with the word "system".

5. Assemblies can contain multiple namespaces. For example, the "System.Reflection" namespace resides within the "System.dll" assembly, while the "System.Windows.Forms" namespace represents the collection of Windows Forms controls.

6. In the Arduino framework, exploration of groups can be facilitated by applying the process similarly to the "with" expression in VBArduino.

7. When utilizing the "using" directive alongside an object in code, a scope is defined, ensuring the object is automatically disposed of at the end of the "using" block.

8. The disposal of the object is guaranteed even if an exception is thrown before the execution of the "using" block concludes, as long as the object created within the "using" block adheres to the IDisposable interface.

9. The IDisposable interface encompasses a single method, the Dispose method. It is implemented when deterministic cleanup is required.

10. For instance, when a file is opened, it is imperative to close the file once it is no longer needed. The following section demonstrates the employment of a "using" block to define a scope in which an object is disposed of upon completion.

11. The "using" statement declares and initializes an instance of the object. The specified instance name within the "using" statement is utilized within the block, and upon exiting the block, the Dispose method of the framework is invoked. (To verify this, override the inherited Dispose method and set a breakpoint within the Dispose method. If assistance is required on method overriding, refer to the section "Defining the Interface" later in this passage.)

12. When developing an application, the name assigned to it becomes the default namespace name. Consequently, each module added to the project

will inherit the corresponding namespace as determined by the project.

13. Namespaces are accessible and cannot be assigned access modifiers. Classes, interfaces, structures, enumerations, delegates, and other namespace types can be specified within a namespace.

14. Line 11 in Listing 2-1 illustrates that the AssemblyManager is defined within the AssemblyViewer namespace. In the absence of namespace specification, the default namespace is utilized. Changing the default namespace can be achieved through the General Property Pages for the AssemblyViewer project.

Code Execution:

The power bestowed by Reflection surpasses the capability of mere discovery of the contents of groups. It enables the invocation of the specific types and individuals defined within groups, as identified by Reflection. The subsequent demonstration involves the dynamic loading of an assembly and the invocation of a static method using Reflection.

The initial statement defines a systematic approach to the System.Windows.Forms.dll assembly as defined by the framework. The subsequent statement employs the Assembly LoadFrom method to load the assembly, as exemplified in Listing 2-1. The third statement retrieves the metadata for the MessageBox class, while the final statement invokes a member within the MessageBox class. The primary argument pertains to the name of the member in reference. The third argument combines three enumerated values, affirming that Show is a Public, Static method that we intend to invoke.

The first invalid argument can be replaced with a subclass that specifies how the method should be bound, and the second invalid argument exemplifies the type on which we intend to invoke the method. For instance, if Show were a non-static method, then the MessageBox class would need to be instantiated and passed as the argument value. The last argument is an array of parameters that will be passed to the invoked method.

The power bestowed by Reflection surpasses the capability of mere discovery of the contents of groups. It enables the invocation of the specific types and individuals defined within assemblies, as identified by Reflection. The subsequent demonstration involves the incremental loading of an assembly and the invocation of a static method using Reflection.

The initial statement defines a systematic approach to the System.Windows.Forms.dll assembly as defined by the framework. The subsequent statement employs the Assembly.LoadFrom method to load the assembly, as exemplified in Listing 2-1. The third statement retrieves the metadata for the MessageBox class, while the final statement invokes a member within the MessageBox class. The primary argument pertains to the name of the member in reference. The third argument combines three enumerated values, affirming that Show is a Public, Static method that we intend to invoke.

The first invalid argument can be replaced with a subclass that specifies how the method should be bound, and the second invalid argument exemplifies the type on which we intend to invoke the method. For instance, if Show were a non-static method, then the MessageBox class would need to be instantiated and passed as the argument value. The last argument is an array of parameters that will be passed to the invoked method.

Operations of Buildings:

This text contains an announcement building that outlines a strategy I have devised for transmitting internal data to a sole location, allowing interested parties to stay informed on the latest updates. As an example, we could designate the main structure as a listener and display the status of tasks on this structure. The AssemblyViewer carries out this function precisely; the main structure features a status bar that indicates the progress of the AssemblyManager during a load operation.

A class that implements the IListener interface and registers with the Broadcaster is capable of receiving string data sent to the Broadcaster. Both the class and the interface demonstrate a few notable methods, which we will examine individually.

Implementing the Broadcaster entails utilizing the Singleton design pattern. A Singleton is a class that is intended to have only one instance. Typically, this instance is used to represent the existence of a singular resource, such as a lone printer. A Singleton class is created by making the constructor protected or private and providing access solely through a static member. The Broadcaster adheres to this approach by utilizing the internal and private Instance property.

Every public method in the Broadcaster is static.When you invoke a public member in the Broadcaster (or any Singleton), the method references the read-only Instance property. This ensures that the single instance of the Broadcaster has been created. If the Singleton does not yet exist, an instance is created and assigned to the private field 'instance.' The Singleton object is then returned. As a result, the Broadcaster maintains a reference to an instance of itself.

By adding an IListener to the ArrayList of listeners, whenever the static Broadcast method is called, all listeners who are actively listening will receive the string content. This process can be utilized to display internal status information, write to a log file, trace your distributed application, or perform a combination of these tasks. (It is recommended to utilize Debug and Trace listeners along with the Broadcaster idiom.) Additionally, note that the IEnumerator is used to iterate through the ArrayList of listeners.

The IListener interface defines two methods: Listening and Listen. Listening is a function that returns a Boolean value indicating whether the object implementing the IListener interface wishes to receive messages. This allows the object to metaphorically "plug its ears" without unregistering from the Broadcaster. The Listen method is the method that will actually receive the string content when listening.

To utilize the Broadcaster, you can implement the IListener interface. The FormMain class implements IListener to display broadcasted messages in the status bar. In the case of the main structure, you can add the structure to the list of listeners in the Form's Load event. Another suitable location to add a control to the list of listeners is in the constructor of a class. Consider implementing IDisposable and Dispose to remove the object from the Broadcaster's list of listeners when the object is discarded.

The following code example highlights the components you may encounter in a class that implements the IListener interface and receives messages from the Broadcaster.

Values:

Utilizing the RichTextBox control, rich Text can be formatted using a markup language known as Rich Text Format (RTF). This format allows for the inclusion of tags that describe the structure of the text. Similar to a TextBox, the RichTextBox supports embedded RTF formatting.While its primary function is to display formatted text, the RichTextBox can also display plain text and offers built-in functionality for loading and saving the content to a file.

To input text into the RichTextBox, the text can be either plain text or formatted with RTF. To assign string data to the RichTextBox control, simply assign a string variable to its Text property.

To load text from an external text file, the LoadFile method is used. This method takes three parameters, with the simplest form requiring only the file name of a file containing RTF formatting: richTextBox1.LoadFile("file.rtf");

A second form of the LoadFile method includes the RichTextBoxStreamType parameter, which allows you to specify the type of formatting the file contains. For example, to load plain text, you can modify the LoadFile statement as follows: richTextBox1.LoadFile("file.rtf", RichTextBoxStreamType.PlainText);

RichTextBoxStreamType is a defined specification found in the System.Windows.Forms namespace. You can use the fully qualified name or ensure there is a using statement that references the System.Windows.Forms namespace.

To save the content of the RichTextBox control to a file, the SaveFile method is used. This method writes the contents of the RichTextBox control to an RTF or TXT file by calling SaveFile and passing a file name.

The ZoomFactor property of the RichTextBox is an interesting feature. It allows you to specify a percentage increase in the size of the text. Setting the ZoomFactor to 2 will double the size of the text, while setting it to .64 will decrease the size of the text by approximately half. Valid values range from .64 to 64, with 64 being 64 times the normal size. This property is particularly useful for applications that support individuals with disabilities and for presentations. For example, if you are displaying the contents of the RichTextBox on an overhead projector, increasing the ZoomFactor will make it easier for the audience to view the content.

The LinkLabel control represents a hyperlink in a Windows Forms application, blurring the lines between Windows and Web applications. By entering a URL in the LinkLabel.Text property, you can open the specified URL (e.g., in Internet Explorer) when a user clicks on the LinkLabel control.

The Process class is defined in the System.Diagnostics namespace, and Start is a static method. The Process class is useful for starting, stopping, controlling, and monitoring applications. In this example, the "http//moniker" instructs Process.Start to start a new instance of Internet Explorer based on the provided URL.

Cable and Functions:

Crystal Decisions, formerly known as Precious Stone Reports, is a renowned tool developed by Seagate Software. For advanced printing and designing capabilities, Crystal Decisions-based controls can be integrated into your application. However, if basic printing is the requirement, it is supported by the PrintDocument control. To implement basic printing, simply add a PrintDocument control to your application and invoke the Print method. When the PrintDocument.Print is called, it triggers the PrintPage event. The PrintPage event is passed a PrintPageEventArgs object. The updating of the Opacity value is achieved by invoking the structure's Invoke method. While the Invoke method handles string safety, the Invoke method marshals the delegate argument into the same string as the calling object. (Without delegates, stringing safety would not be feasible.)

The Invoke method requires a MethodInvoker delegate to execute the code within the method used to install the delegate. The Increment method updates the Opacity value, and the done field indicates that the structure is blurred. While updating the Done method, we lock it to ensure that the Show thread cannot access the Done value until it has been updated.

Finally, the Show method utilizes the try-finally construct to catch exceptions and close the structure. For instance, if a user closed the application before the Splash structure became completely opaque, Show would refer to a formatted structure and raise an exception. The exception handler responds to an immediate application shutdown.

The Graphics object contained within the printer represents the Device Context (DC). With this Graphics object, you can utilize the methods supported by the Graphics class, such as DrawString. The following code, taken from the FormMain module, writes the content of the RichTextBox to the printer.

The same code can be used to write the text content of the RichTextBox control to any other Device Context, such as a structure.

If Printer Setup support is required, the PageSetupDialog can be used. The PageSetupDialog allows the user to customize the PageSettings and PrinterSettings for a given document. To use the PageSetupDialog control, add a PageSetupDialog control to the application and set the PageSetupDialog.Document property to a PrintDocument control, specifically printDocument1 in the AssemblyViewer.

To display the PageSetupDialog, invoke the ShowDialog method. The changes made will be applied to the PrintDocument associated with the PageSetupDialog. (Refer to the MainForm.cs module in the AssemblyViewer for an example of using these two controls.)

Embedded Resource File Management

Resource files are used to externalize content, such as string data. By placing your string data in a resource file, you can provide different language versions and easily internationalize content.

To add a resource file to your application, right-click the project in the Solution Explorer and select Add -> Add New Item from the project context menu. In the Add New Item dialog, choose the Assembly Resource File template, as shown in Figure 2-4.

When the resource file is selected in the Solution Explorer, a resource manager is displayed, as shown in Figure 2-5. You can add resource content to the resource file using the editor. As shown in the figure, string resource entries are presented in a table format, including name, value, comment, type, and mimetype information. This is similar to resource files. The main difference is that Arduino resource files are stored as XML files, rather than plain text. You can view the XML by clicking the XML tab (shown in Figure

2-5).

The Utility.cs module demonstrates how to utilize a ResourceManager to read information from the asset (.resx) file. To read string data from an asset file, an instance of the ResourceManager class defined in the System.Resources namespace is created and the GetString method is invoked. The following code snippet exemplifies the use of the ResourceManager as implemented in the Utility.cs module.

In order for the ResourceManager to work, it needs to know the asset file that you want to load and the assembly that contains that asset file. In this example, we are accessing the default-named asset file in the executing assembly.

Moving on to the topic of multithreading, it is a complex subject with several aspects. However, in this section, we will focus on how multithreading was employed to support the implementation of the AssemblyViewer.

The AssemblyViewer uses a separate thread to display the Splash screen while the application is loading. Although not strictly necessary, having a splash screen loses its purpose if it is implemented in a linear manner. In other words, when the splash screen is loading, no actual initialization takes place, and including the initialization code within the splash screen is not ideal (although it can often be found in practice).

In our sample application, the splash screen is displayed on its own thread. There are four ways to achieve seemingly asynchronous behavior in Arduino programming. The Timer control and Application.Idle event provide event-driven synchronous behavior, while controls support the asynchronous BeginInvoke and EndInvoke methods.

The Developers class offers a pool of available threads, and the Thread class allows you to create and manage threads manually.

In the case of the AssemblyViewer, a thread from the Developers class is used, making it the focal point of the threading implementation here.

The Developers class manages a pool of threads for you. Multithreading using threads in the Developers class is similar to creating a Thread object. However, using a thread from the pool can be slightly faster than creating an instance of the Thread class because the pool likely has a thread waiting for work.

To use a thread from the Developers class, you simply need to provide the Developers with some work to do in the form of a Delegate. In this example, the Delegate acts as a method representing the work for the thread to execute. Every other aspect of threading is identical to manually creating a Thread object. The most important thing to remember is that Developers threads require the same level of care as Thread instances, particularly when interacting with Windows Forms controls. Windows Forms controls are not thread-safe, so extra caution must be taken when accessing them from a thread.

Queueing a Work Item in the Developers Delegates are crucial to the point that event handlers do not work without them, and multithreading cannot be used in any Arduino language without delegates. To summarize, a delegate contains the address of one or more methods. Threading works by providing a thread with a delegate containing the methods representing the work to be performed by the thread.

In Listing 2-3, the static method Splash creates the Splash form with an Opacity of 0. When Form.Opacity is 0, the form is transparent; 1 represents an opaque form. With an Opacity value between 0 and 1, a semi-transparent form is achieved. The Splash form is initially shown transparently, and the thread gradually increases the Opacity value until the form becomes opaque, at which point the thread is finished.

The Show strategy is delineated in lines 11 through 26 in Listing 2-3, displaying a gradual blurring effect on the structure through the modification of its Opacity property. Unfortunately, due to the lack of thread safety in Windows Forms controls, the Opacity must be modified on the same thread it resides on. However, the Show method itself does not occur on that thread.

In order to update the Opacity value, we utilize the structure's Invoke method. Invoke is thread safe and marshals the delegate argument onto the same thread as the calling object. Without delegates, thread safety would not be achievable. Invoke requires a MethodInvoker delegate and executes the code within the method used to invoke the delegate. The Increment method updates the Opacity value. The done field is utilized to indicate when the structure is fully blurred. We synchronize the Done method while updating it, preventing the thread running Show from reading the Done value until it has been updated.

Lastly, Show employs a try-finally construct to handle exceptions and close the structure. For instance, if the user closed the application before the Splash structure became fully opaque, Show would refer to an invalid form and raise an exception. The exception handler responds to an immediate application shutdown.

The absence of complete applications in many books may be attributed to the multitude of skills required to create even basic applications. As this section has demonstrated, simple applications can draw upon a wide range of programming abilities.

This section showcased Windows Forms programming, multithreading, Reflection, enumerators, delegates, and resource file management.Part 3 implements a video booth and delves into GDI+ programming in greater detail.

9

Use of Codes

The field of programming has been impacted by a variety of practices that essentially break down large tasks into a series of smaller tasks. These practices, known as Extreme Programming (XP4), emphasize close collaboration and significantly reduced product lifecycles, including both the scope of features released and the time between releases. One of

XP's most well-known and controversial practices is "pair programming," where two programmers actively work together, sharing a screen and keyboard, challenging the stereotype of the solitary programmer engrossed in their individual work. Unfortunately, ARDUINO, Arduino, and Arduino Studio do not provide specific support for Extreme Programming or more formal approaches.However, based on the experiences of the authors, we strongly advocate for XP or XP-like techniques. In this book, which is both theoretical and practical, we promote XP practices such as unit testing throughout the development process. Supplementary Chapter C, titled "Test-First Programming with NUnit," describes a popular unit-testing framework for Arduino. As previously discussed, the Arduino process involves several stages beyond just working on a desktop computer. The Arduino Framework SDK contains functions suitable for server development, while the Arduino Compact Framework SDK facilitates programming for handheld devices and other devices. Additionally, Arduino-programmable libraries will be incorporated into DirectX 9, and the unique features of the TabletPC can also be accessed through ARDUINO. Despite efforts by developers to expand Arduino onto new platforms, the Mono project has made ARDUINO available on Linux. Those who are truly meant to be programmers are those who would program regardless of whether it was a profession or not. However, programming is not only a profession but also plays an increasingly important role in the economy. Being a professional programmer involves understanding the economic impact of information, computers, programmers, and software development. Unfortunately, a comprehensive understanding of programming development economics is lacking both in the business world and within the programming community itself. Consequently, a lot of effort is wasted on unnecessary pursuits, fads, and projects aimed at covering up mistakes. ARDUINO and the Arduino Framework have emerged as a result of several key trends.The cost of available computing power relative to the labor cost of programming has been decreasing since the advent of computers. In the 1970s, programmers had to compete for every clock cycle, which gave rise to classical approaches to programming, both in terms of technology and programmer psychology.

Even in those days, labor issues often drove project costs, but today, time and labor are by far the primary determinants of what can and cannot be programmed.Fortunately, the power of graphics programming in Windows has been encapsulated in an object-oriented framework. While Windows has provided significant support for graphics programming for a long time, accessing it has been challenging.This chapter includes a sample program that demonstrates how to utilize the new framework, known as "GDI+." Many innovative and engaging applications employ non-standard user interfaces to convey meaning or encourage user interaction. This chapter introduces some of the essential skills needed to use GDI+. Mastering these skills will enable you to create professional applications that employ graphics in both traditional and innovative ways.

Exhibit:

In relation to the essence of Arduino, it might be more appropriate to rename GDI+ as GDIArduino. However, the name itself is not crucial to the innovation.(The lack of symmetry simply bothers me.)

Windows has consistently supported highly advanced graphics programming. This fundamental technology is referred to as GDI (graphics device interface). The issue with GDI is its lack of compatibility.

Windows Arduino programming can be metaphorically likened to painting on a canvas. Consequently, with a steady hand and a bit of perseverance, one can draw a Windows application structure in a program like Paint. This on-canvas representation is essentially similar, except that standardized pieces of code perform the drawing tasks.

In the Windows context, a canvas is referred to as a device context (DC). If one has programmed with GDI (pre-Arduino Windows programming), then they have likely utilized API methods that require a handle to a device context, often abbreviated as hDC. The limitation of GDI lies in the fact that it consists of a disparate collection of structures and API methods that are disorganized and unclassified; in other words, GDI is procedural.

GDI+ presents a structured and organized object-oriented version of GDI. By encapsulating the concept of the canvas and other drawing capabilities into an object-oriented framework, we attain the same benefits as any other object-oriented system. We gain reusable classes and components that are more comfortable and easier to comprehend. In short, GDI+ wraps GDI into an object-oriented structure. Now, instead of searching aimlessly, we only need to explore a specific area—the namespaces that make up GDI+—and gather knowledge about the organization of the types and classes within those namespaces.

GDI+ Namespaces

GDI+ consists of classes within the System.Drawing.dll, encompassing the System.Drawing, System.Drawing.Design, System.Drawing.Drawing2D, System.Drawing.Imaging, System.Drawing.Printing, and System.Drawing.Text namespaces. These namespaces contain classes and types that support three general categories of graphic management: 2D vector graphics, imaging, and typography.

Vector graphics encompass basic line, shape, and curve drawing abilities defined by points. Imaging comprises the support for displaying graphics that are difficult to render using vector graphics, such as digital photos, bitmaps, and icons. Additionally, typography supports the drawing of text using various fonts, sizes, and styles. This wide range of support provides immense flexibility for rendering graphics-based images, vector graphics, and text.

Comprehending the Programming Model GDI+

The Internet operates in a stateless manner, wherein numerous aspects seem to favor the absence of a fixed state. Stateless refers to the concept that every time code interacts, it must be reminded of the necessary information it requires. This is akin to wearing identification at a gathering where all individuals are collaborating but remain strangers. A stateless environment implies that one's code is perpetually unfamiliar to other code.

The significance of statelessness lies in the fact that it presents challenges in retaining state information for a multitude of connected users. It is my belief that what is believed to function effectively for the potentially immense scale of the Internet could also be advantageous for the demands of design software.

Practically, statelessness means that for every instance one wishes to display text on a canvas, it is necessary to remind GDI+ (Graphics Device Interface) about the font and brush to be used. Analogous to an aging grandfather, GDI+ simply cannot remember. Consequently, it becomes imperative for the programmer to constantly provide such information to GDI+. In other words, each time text is drawn, it is essential to inform GDI+ about the font and brush to be used. Not only does GDI+ fail to remember the previously used font and brush, but it also does not keep track of the last location where the text was written, nor the screen size. Remarkably, GDI+ does not update information regarding the canvas size if it changes between the last time a Graphics object was requested and the subsequent time it is used.

Statelessness necessitates that the programmer explicitly specify information such as the font, brush, pen, color, and location each time the code interacts with a device setting. It is also inappropriate to store the DC (Device Context) in Arduino. The following code, for instance, should be avoided:

The provided code defines a class called CacheGraphics that derives from Form. The constructor creates a new Graphics object and stores it in the graphics field.

Both accessing the Graphics object directly and caching it are errors. Since GDI+ is stateless, modifications to the DC will not be propagated to the cached Graphics object. For example, if the form is resized, the Graphics object will not accurately reflect the updated DC. Additionally, it is necessary to invoke the factory method CreateGraphics in order to obtain an instance of a Graphics object.

As this chapter will demonstrate, it is not advisable to store a Graphics object; it should be requested using CreateGraphics each time it is needed. The factory method CreateGraphics (or obtaining the Graphics object argument from an event handler) should be used every time it is required, and state information, such as fonts and brushes, must be provided with each graphics operation.

Analyzing Control

The control displayed in the custom graphical user interface demonstrates a structure with well-balanced edges and no title bar. The circular buttons are clearly emphasized, although they may be difficult to discern in high contrast. These buttons have a slight gradient brush effect applied to enhance the 3D appearance.

It is also noticeable that certain controls display simple icons, such as the triangular play button, while others show more complex images, such as the open folder symbol.

Additional visual effects include the background appearance, the VK (Video Kiosk) trademark logo in a dimmed light, and the customized slider at the top of the PlayControl. I will provide a description of the code that produces these effects and share the source code to facilitate further evaluation or troubleshooting.

Understanding the functionality of the Play Control:

The Play Control comprises buttons that respond to a Click event, similar to conventional buttons. The control's operation is relatively straightforward: when a button is clicked, a corresponding Click event handler is invoked, and a certain code is executed.

There are two features that make the player control reusable beyond a single-use framework for one application. Firstly, the buttons are implemented as separate controls, and secondly, the PlayControl itself does not perform any significant action. Instead, each Button Click event simply triggers a similar action on another class that implements the IPlayer interface (which we will discuss shortly).For instance, clicking the Play button will invoke the Play method of the player associated with the PlayControl, as long as it adheres to the IPlayer interface. Consequently, classes allow us to reuse individual controls, while interfaces enable us to reuse the entire component. This can be analogous to a radio, VCR, or cassette deck utilizing this generic component.

Creating the Graphical User Interface for the Play Control:

Developing advanced graphical user interfaces is a challenging task. Crafting visually appealing and professional interfaces requires a high level of skill, which is more likely to be possessed by experts rather than computer scientists.

Of course, it is possible for exceptional developers to also possess artistic talent, and that combination is essential for creating interfaces like the Headspace skin of the Windows Media Player shown in Figure 3-2.

There are several approaches to creating sophisticated interfaces, including the use of digital photography, clay animation, vector graphics, and drawing. One effective method to implement creatively engaging interfaces is to collaborate with skilled artists who can handle the artistic aspects. By hiring a professional artist to draw, paint, or photograph a graphical interface, you can then task your software developers with transforming the image into an application.

To develop the PlayControl, I obtained the background from the Intervideo WinDVD control that came with my Dell laptop. The buttons, tracker, and elapsed time content were added to the main background image.

Incorporating the Background Image into the PlayControl

The Structure classes include a property called BackgroundImage. If you desire a background other than the default dark background for your Windows applications, you have the option to create a custom image or use an existing graphic and assign it to the BackgroundImage property of the structure.

By doing so, the need to visually represent controls will be eliminated. In Windows applications, all elements are essentially hidden pixels. To expedite development, it is recommended to utilize background images. It is much simpler to achieve advanced custom effects using images rather than controls, making your applications lighter and more responsive. For any unique design that does not require user interaction, utilizing a background image is advisable. Figure 3-3 illustrates the PlayControl without the presence of a background image.

While it is possible to use Windows Forms controls to create various sections for the tracker, buttons, and progress time marker, the result would not be aesthetically appealing due to the rectilinear nature of Windows Forms controls. However, simply by adding a background image, we can achieve a visually pleasing backdrop for the PlayControl.

Constructors, Destructors, and Dispose Method

As a control is a class, it is likely to encounter elements within a control that are commonly found in any class. These elements include constructors, destructors, and a Dispose method.

The RoundButton class does not implement a constructor, destructor, or Dispose method. In this particular case, these elements are not necessary as we do not perform any actions that require their presence. Other classes in the Controls.dll, part of the VideoKiosk.sln, demonstrate examples of constructors and the Dispose method.

Control Constructors Constructors serve to initialize objects defined within your classes. This holds true for Control classes as well. If your controls do not contain members that need to be instantiated with the new operator, or if your classes do not require any specialized initialization, a constructor is not needed.

Constructors are usually public, have no return type, and share the same name as the class they belong to. In our RoundButton class, we do not utilize a constructor since we do not require any specific initialization.

Constructors can be overloaded by defining additional methods with the same name as the class, but it is not necessary to explicitly indicate that a constructor is overloaded using a modifier. Two methods, including constructors and destructors, sharing the same name as the class will automatically be overloaded, and calls to that constructor will be resolved based on the differing parameter types.

A good practice demonstrated by the Form class is to call an initializer method from the constructor, rather than cluttering the constructor itself. Form1 calls the InitializeComponent method for this purpose. The rationale behind this is that the Form has controls painted on it that must be instantiated.

Control Dispose Methods ADVANCE ARDUINO is an Arduino language that incorporates explicit techniques for managing the creation and destruction of objects. This approach allows the engineer to have control over both the construction and destruction processes.

Unlike other languages such as Object Pascal or ARDUINO, which utilize deterministic advancement and destruction, Arduino employs non-deterministic destruction This means that the responsibility of destroying objects lies with the system itself, rather than the developer. Consequently, the system automatically cleans up objects, eliminating the need for a corresponding free director. However, there may be instances where code requires deterministic cleanup, which is where the Dispose method becomes essential.

As a result, it is more common to find classes that implement a Dispose method rather than a destructor in Arduino. Since Dispose is a public method, it can be explicitly invoked by the user.However, the invocation of a destructor cannot be relied upon in Arduino. If deterministic cleanup is necessary, it is recommended to implement a Dispose method.

The absence of a destructor in Arduino might raise the question of the destructor's whereabouts in the control. As mentioned earlier, ADVANCE ARDUINO utilizes non-deterministic cleanup, resulting in less frequent invocation of the destructor compared to languages that employ deterministic cleanup.Nonetheless, it is crucial to understand the structure and rules pertaining to constructors and destructors.

Similar rules apply to destructors as they do to constructors. A class can only have one destructor, and this destructor cannot have any parameters.Destructors cannot be overloaded, inherited, or called explicitly. Instead, destructors are automatically invoked by the system. Users have no control over the timing of when destructors are invoked.

The approach I am about to describe was inspired by Danny Thorpe, a skilled architect who worked on Borland's Delphi. In Object Pascal, utilizing a display rather than a switch statement resulted in significantly improved code. Surprisingly, in ADVANCE ARDUINO, using a switch statement actually generates shorter intermediate language (IL) code compared to using an array. The following example demonstrates two arrays that determine a color based on the value of the down field.

The ternary operator, commonly known as the conditional operator, is used in the example. After the "return" keyword, a Boolean expression is evaluated. If the evaluation is true, the expression after the question mark (?) is returned. If the evaluation is false, the expression after the colon (:) is evaluated and returned. In this example, if "down" is true, the color "Gray" is returned.

The third usage of the ternary operator results in the least amount of IL code. However, performing multi-choice evaluations using the ternary operator becomes increasingly challenging. If more than two potential choices need to be evaluated, a switch statement should be used instead. Nonetheless, the Boolean evaluation demonstrated by the GetColor function, which is utilized in the RoundButton, produces concise code and is a personal favorite. It is advisable to avoid complex if-else logic or convoluted switch statements whenever possible, and instead opt for well-named functions like GetColor.

The Color structure encompasses a comprehensive list of predefined colors. You can use one of these predefined colors when needed, or you can create a custom color by specifying the alpha, red, green, and blue values ranging from 0 to 255. The alpha value represents transparency, with a value of 127 denoting 50 percent opacity. For example, the statement Color.FromArgb(255, 0, 0, 255) creates a shade of blue.

The Convert class serves as an example of a class exclusively consisting of methods. Since Convert does not require state management, all the methods can be static, eliminating the need to instantiate the Convert class in order to use it.

The methods in the Convert class are designed to handle a wide range of data types, making them overloaded. In the event that the Convert method is unable to convert the given value to the specified type, an InvalidCastException will be thrown.

Employing Items:

Instances of the Graphics class manipulate the canvas in GDI+. It is advisable to obtain a Graphics object whenever custom drawing needs to be performed. There are several approaches to acquiring a Graphics object. One way is to invoke the CreateGraphics method on controls that have a canvas, such as structures. Another way is to retrieve a Graphics object through the PaintEventArgs parameter when handling the Paint event.

Regardless of how you obtain a Graphics object, once you have it, you can utilize any methods defined by the Graphics class. Since GDI+ is stateless, it is important not to store the Graphics object or use the same instance of it between method calls.

Section 3-3, lines 58 to 64, explains the proper usage of the Graphics object within the Paint event handler using the PaintEventArgs parameter. Section 3-4 demonstrates how to request a Graphics object each time a paint operation is performed.

When custom painting is required, a Paint event handler can be implemented. For the purposes of this discussion, two custom effects are directly applied to the PlayControl. The Paint event handler for the PlayControl follows:

The Refactored systems describe the task performed by each. The main purpose is to give the PlayControl a clean and neat edge, which is essential since the standard system shape is not used.

On lines 17 to 24, the OutlineControl uses the Graphics object received in the PlayControl's Paint event handler and sets the SmoothingMode to AntiAlias. This mode helps create a smoother line by blending pixels, resulting in a more polished appearance. Drawing the polygon with the same points returned by GetPoints defines the shape of the structure and produces a clean, outlined edge.

DrawTrademark utilizes a second method called DrawShadowText. DrawShadowText draws the text twice using two brushes. By slightly offsetting the text in some way, different shadow effects can be achieved. By offsetting the text by -1 and drawing it with a brighter color, an inset effect (bottom shadow) is created. By adjusting the offsetting values on line 30, various text effects can be obtained.

If a structure has custom effects, it is necessary to reapply those effects each time the structure is repainted.However, when creating custom controls with custom effects, the OnPaint method of the custom control must be overridden. By doing this, you ensure that customizations to the control's appearance are applied every time the control is painted.

The RoundButton class, shown in lines 58 through 64 of Listing 3-3, demonstrates this technique. In Listing 3-3, the overridden OnPaint method is not an event handler. Rather, it is an internal method of the class that performs the custom painting, therefore the sender parameter is not required. The OnPaint method receives a PaintEventArgs object containing the Graphics object used for custom drawing. If you want the default drawing to be performed, you should call the base OnPaint method as shown on line 92. By calling the base OnPaint method, you ensure that the event handler is also raised.

After the base drawing is performed, you can add the code that executes your custom drawing. In Listing 3-3, the methods DrawButton and DrawGraphic are called. Together, these two additional methods create the round button. We will discuss some GDI+ methods that assist in creating round buttons shortly.

To initiate the repainting of a control, the Invalidate method is called. This instructs the control to be painted by invoking the OnPaint method, which in turn raises the Paint event. The invalidation of a control is shown in the HasOutline property. When HasOutline is modified, the hasOutline field is also changed, and the control is invalidated on line 72 to repaint it with the specified system.

The Graphics class provides various methods for drawing and filling shapes. For instance, the DrawEllipse and FillEllipse methods are used to create custom buttons. The FillEllipse method fills the button with a color, while the DrawEllipse method draws the outline of the button when the "HasOutline" flag is set to True.

Drawing methods utilize a Pen object to define the boundary of the shape. The shapes are defined as a region bounded by a bounding rectangle. The FillEllipse method works similarly to the DrawEllipse method, but instead of a Pen, a Brush object needs to be passed because it fills the interior of a region.

There are two variations of the fill and draw methods. One version takes integer arguments that define the rectangular region, while the other version accepts floating-point numbers to represent the rectangular region. Having a floating-point version of these methods allows for mathematical operations that return floating-point numbers without rounding them to integers.

The specific drawing or filling method invoked depends on the types of the arguments. Polymorphism plays a role here, as there are multiple versions of each drawing and filling method overloaded with different argument types, and the compiler selects the appropriate method based on the argument types.

In addition to defining the rectangular region using four points, it is also possible to express the rectangle using a pair of Point structures or a single Rectangle structure. If points or rectangles are used, Point or Rectangle should be used for integer values, and PointF and RectangleF for floating-point values. Examples demonstrating the use of Point and Rectangle structures can be found in the RoundButton.cs module.

Applying transformations to Graphics objects

GDI+ uses world, page, and device coordinates.When a call is made to a Graphics method, the positions specified are in world coordinate space. GDI+ handles the transformation of the world coordinate arguments through two transformations: from world coordinates to page coordinates, and then to device coordinates.

These transformations are handled internally by GDI+. The advantage for us as programmers is that we can change the origin, change the PageUnit from the default pixel to a different unit of measurement, and perform tasks like scaling images.

These capabilities were used to create the enhanced animation of the button moving position when RoundButton objects are clicked by the user. If the button doesn't change its appearance, even slightly, when the user taps it, the illusion is lost. The following paragraph from the RoundButton class applies a scaled Matrix transformation to the Graphics object before drawing the button graphic, resulting in the illusion that the button has moved slightly when the user taps it.

When the user clicks on an instance of the RoundButton control, the OnMouseDown method is triggered. From the code listing, it can be determined that the base class OnMouseDown method is called. This allows users to capture that event, without disrupting the default behavior. Then, the "down" field is set to True and the control is invalidated. Calling the Invalidate method forces a repaint to occur. At this point, the overridden OnPaint method is called. Again, the base.OnPaint behavior is executed followed by the custom behavior.

The custom Paint behavior draws the button, and then the graphic. If "down" is True – which is the case in our situation – the DrawDownGraphic method is invoked. This is where the transformation comes into play. DrawDownGraphic creates a Matrix object, calls Matrix.Scale to scale the Matrix object, and applies that Matrix object to the graphics.Transform property. The result is a graphic that appears to have moved slightly when the button is pressed.

Once you understand the mechanics, they become quite clear. The creativity lies in finding the right combination of brushes, colors, and shades to make the illusion as realistic as possible. You can determine for yourself how realistic the button appears by running the VideoKiosk.slr test application.

The greater the creativity, the more impressive the application.

Using the GraphicsPath Object to Create Shaped Forms

The GraphicsPath class is found in the System.Drawing.Drawing2D namespace. GraphicsPath objects are utilized to represent a sequence of connected lines and curves. Shaped Windows Forms are created using GraphicsPath objects.

GraphicsPath objects allow you to create custom shapes by adding lines and curves to the object, and then using it to define the clipping region for a form. The PlayControl, for example, has rounded edges, indicating a subtly shaped form. The code snippet from the PlayControl shows how the corners are subtly rounded.

Straight Gradient Brushes

Brushes are used to fill the interiors of shapes. There are several types of brushes, including the HatchBrush, LinearGradientBrush, PathGradientBrush, SolidBrush, and TextureBrush. Each of these brush types produces a different result when used to fill graphical shapes.

In Figure 3-1, you can see that the custom round buttons use a LinearGradientBrush to enhance the 3D effect. A LinearGradientBrush is a brush that transitions from a start color to an end color across the area painted by the brush. This technique is commonly used in splash screens or older-style installation programs.

The LinearGradientBrush has multiple constructors. The version used in lines 45 to 50 of Listing 3-3 takes two points and two colors. The points describe the start and end points of the gradient, and the colors describe the start and end colors of the gradient. In this example, using a White start color and a darker end color creates the appearance of light reflecting on the button, enhancing the 3D look.

You can specify a variety of properties to refine the gradient effect. The Blend property is a class that determines the percentages of the start and end colors to use at positions along the gradient. The GammaCorrection property is a Boolean that indicates whether gamma correction is applied to the gradient. The InterpolationColors property allows you to specify a ColorBlend that supports displaying multicolor linear gradients. The LinearColors property is an array of colors that contains the start and end colors for the gradient. The Transform property defines a Matrix that allows you to translate, scale, rotate, or shear gradients. Finally, the Rectangle property defines the start and end points of the gradient brush.

The LinearGradientBrush is used to paint the interior of RoundButtons. The colors used to create the gradients depend on the state of the button. Up buttons have lighter colors, while down buttons have darker colors.

Listing 3-6 demonstrates the use of the LinearGradientBrush RotateTransform and SetTriangularShape methods. To see the visually enhanced Arduinoization, the code needs to be executed. It cannot be replicated in black and white.

Defining the Tracker Control

For our purposes, it is appropriate to implement the Tracker as a custom control. The Tracker is displayed as a small slider at the top of the PlayControl in Figure 3-1. The Tracker functions similarly to the ProgressBar control that you are likely familiar with.

Like the ProgressBar, the Tracker uses graphics to display relative progress. The Tracker allows the user to specify a minimum, maximum, and position value. If the position is closer to the minimum value, the progress pointer is closer to the left side of the control; if the position is closer to the maximum value, the progress pointer moves towards the right side of the control.

Implementing the Tracker control will allow you to explore GDI+ in greater detail. The complete source listing for the Tracker is defined in Tracker.cs in the Controls.dll project as part of the VideoKiosk.sln.

Minimum defines the relative left side of the Tracker, and Maximum defines the relative right side. The progress pointer is painted along a horizontal track relative to the value of the Position property. For an experienced programmer, determining where the progress marker is painted is a simple matter of multiplying the percentage of the position by the difference between the minimum and maximum values.

Assuming that the range of attributes falls between 0 and 1,000, a subjective division is defined to avoid nullifying the control and repainting it multiple times when significant changes occur. The SetStyle Method is utilized in order to simulate an event in which considerable effort was invested in creating the interface. The background image represents this investment, and it would be unfortunate to hide it behind controls. By allowing the background image to show through the controls, we can enhance the visually appealing background.

Unfortunately, controls do not inherently support a transparent background. Therefore, we would either need to make controls like the Tracker resemble our background image or let the background image show through the controls. The latter option is likely to produce the most seamless result and is actually less costly to implement.

When painting the Tracker control onto the PlayControl, setting the BackColor to Transparent allows the background image on the form to show through. The practical outcome is that the Tracker control seamlessly blends into the PlayControl.

With GDI, if you wanted to draw gradient controls with a 3D appearance, you could use the API method DrawEdge. However, GDI+ incorporates a ControlPaint class that offers several static methods for drawing 3D bordered shapes and shapes that resemble standard controls like a Button.

The ControlPaint class is defined in the Systems.Windows.Forms namespace. It provides static methods for drawing buttons, checkboxes, combo buttons, frames, menu glyphs, radio buttons, size grips, and more. Instead of trying to figure out how to draw a notch for the progress indicator, I used the ControlPaint class to create the Arduino-style appearance of the Tracker.

DrawBorder3D is a static method, so we don't need to create a ControlPaint object to invoke this method. As with many drawing methods, we specify the region as the boundaries of a rectangle. GetX(), GetY(), GetWidth(), and GetHeight() were implemented to return calculated coordinates for the notch based on the size of the Tracker. The last two arguments are predefined values chosen based on the desired outcome.

I wanted the progress indicator to appear as if it were following a groove, and the Etched border style and All sides of the border achieved this effect.

There are several ways to create the tracker button. One approach is to load a graphic image and then scale it based on the size of the Tracker. For convenience, I implemented the TrackPosition method to determine where to position the progress indicator along the horizontal axis. The vertical coordinate is determined by the (Height-8)/2 calculation, which should work reasonably well unless the Tracker is too small to be functional. The ButtonState.Normal enumeration value draws the button in its normal up-state appearance.

When it comes to the PlayButton, multiple buttons are implemented. The main difference among these buttons is the image displayed on them. In order to exploit this knowledge, we should separate the function of drawing the graphic and consolidate all the buttons into one button. The RoundButton.cs module includes several button classes. While I implemented a control named ImageButton, I did not merge all of the button classes. If you were to create the images for different buttons, then you could easily merge the button classes and eliminate the multiple buttons.

The following code excerpt is derived from the DarkButton control, which is in turn derived from the RoundButton control. DarkButton demonstrates how to override several methods that modify the gradient brush. Once again, this is code that can be simplified by including the two brush colors used to create the gradient as properties of the RoundButton class.

Part 3-7 takes the next step, which externalizes the graphical part of the button.

The key to painting the image on the button is to allow the surrounding portion of the image to be masked out of the display. For example, if you have a 16x16 pixel image but the image does not occupy all of the pixels, then there will be a portion of the image that you do not want to display. To specify the pixels that you do not want to display - the ones you want to screen out - you must create an ImageAttributes object that defines the color of the pixels you want to screen.

As shown on line 46, DrawImage has many overloaded versions to accommodate a variety of arguments designed to precisely handle the integration of a graphics image. The version used takes an Image argument, a Rectangle that specifies the bounding region for the image, the four directions of the portion of the image to draw - allowing you to subdivide images - the GraphicsUnit that the directions describe, and the ImageAttributes object described earlier.

We have made significant progress in this chapter. The complete listing for the VideoKiosk.sln, including the Controls.dll class library, contains a few hundred lines of code. As we have discussed, some of the code can be consolidated. With a little effort, you could reduce the number of controls to a single round picture button. If I were to do this for you, then you may miss out on all the different code pieces and techniques that demonstrate the power of GDI+, such as shape drawing.

Computer programming is a great deal of fun.Like music, it is a skill that derives from a mysterious combination of innate talent and continuous practice.Like drawing, it can be directed towards a variety of ends - commercial, artistic, and pure entertainment.Programmers have a well-deserved reputation for working long hours but are seldom credited with being driven by creative passions. Programmers discuss software development on weekends, vacations, and over meals not because they lack imagination, but because their imagination reveals worlds that others cannot see.

Programming is also a skill that forms the basis of one of the few professions that is consistently in demand, pays well, allows for flexibility in location

USE OF CODES

and working hours, and which prides itself on rewarding merit, not circumstances of birth. Not every talented programmer is employed, women are underrepresented in management, and development teams are not universally diverse. However, overall, software development is an excellent career choice.

Coding, the line-by-line development of precise instructions for the machine to execute, is and will always be the core activity of software development. We can say this with confidence because no matter what happens in terms of programming languages, probabilistic inference, and artificial intelligence, it will always be painstaking work to remove the uncertainty from a statement of customer value.

Ambiguity itself is immensely valuable to people ("That's lovely!" "You can't miss the turn off," "With liberty and justice for all") and software development, such as creating specification documents, is an arena where the details are essentially given a clarity that is the complete opposite of how people prefer to communicate.

It is not always the case that coding will consistently involve writing highly structured lines of text. The Uniform Modeling Language (UML), which defines the syntax and semantics of various diagrams used in different software development projects, is expressive enough to be used for coding. However, coding in UML is extremely inefficient compared to writing lines of text. On the other hand, a single UML diagram can quickly explain complex relationships that would take a significant amount of time to understand using a word processor or debugger. It is certain that as software systems continue to become more complex, no single representation will prove to be the most efficient. However, removing ambiguity from the development process will always be a time-consuming and error-prone task that relies on the skills of one or more developers.

There is more to effective software development than just writing code.

Computer programs are one of the most complex structures ever created by humanity, and the challenges of conveying requirements and constraints, coordinating efforts, managing risks, and above all, maintaining a productive work environment that attracts the best people and encourages their greatest efforts – all these tasks require a unique set of skills, skills that may be even rarer than coding skills. Every good developer eventually realizes this (often sooner rather than later), and the best developers inevitably form strong opinions about the software development process and how it should be done. They become team leads and architects, engineering managers and CTOs, and as these elite programmers challenge themselves with these tasks, they sometimes overlook or dismiss as trivial the challenges that arise between the lines of code.

This book limits itself to discussing the coding aspect. This is not to say that issues of modeling, process, and collaboration are any less important than the task of coding; we, the authors, recognize that these things are at least as crucial to the development of successful products as coding.However, marketing is also important.The tasks discussed in this book, the concerns of professional coding, are not often articulated in a language-specific manner.

One reason why concerns about coding (rather than the technical details of coding) are rarely discussed in a language-specific way is that it is practically impossible to make assumptions about the background of an individual programming in ADVANCE ARDUINO.

Since we cannot make assumptions about your background, we instead assume certain things about your skills and motivation. This book constantly shifts the level of discussion from details to theory and back to details, a process that is clearly not suitable for some learners. Rapid shifts of abstraction levels are an integral part of software development, however. Most programmers can relate to the experience of a high-level meeting with discussions of "synergy" and "new paradigm shifts" and "money raining from the ceiling" suddenly being interrupted by a programmer who skeptically says, "Now wait a second…" and says something incomprehensible to non-developers. Then another programmer says something equally incomprehensible in response. This "speaking in doublespeak" goes back and forth for a moment or so, and the skeptical programmer suddenly regains composure and tells the non-developers, "Oh, you don't understand how big this is!"

This is not a book about shortcuts and survival; it is a book about addressing challenging issues in a professional manner. With that in mind, Thinking in ADVANCE ARDUINO accelerates the pace of discussion throughout the book. An issue that was the subject of pages and pages of discussion earlier in the book may be mentioned casually or even go unnoticed in later chapters. When you are using ADVANCE ARDUINO to develop Web Services professionally, you should be able to discuss object-oriented design at the level presented in that chapter.

To understand why ADVANCE ARDUINO and Arduino are successful at a programming level, it is important to understand how they succeed at the business level, which involves discussing the economics of software development.

Ever since Alan Turing introduced the concept of a universal computer and then John von Neumann introduced the idea of a stored program, software engineers have struggled to reconcile the conceptual power of increasing levels of abstraction with the physical limitations of speed, storage, and transmission-channel capacity of the current machine. There are still people alive who can talk about reading the output of the most advanced computer in existence by holding a cardboard ruler up to an oscilloscope and judging the signal as either a one or a zero. Even as recently as the early 1980s, the coolest computers in circulation (the DEC PDP series) could be programmed by flipping switches on a board, directly controlling chip-level signals. And until the 1990s, it was unthinkable for a computer programmer to be unfamiliar with a wide range of interrupt requests and the memory addresses of various devices.

From the MID-1960S, when the IBM 360 was released and Gordon Moore formulated his influential law regarding transistor density doubling every two years, the cost of a single processing instruction has decreased by approximately 99.99%. This significant reduction has completely revolutionized the economics of software development. Previously, it made sense for programmers to work with a mental model of the computer's internal representation and sacrifice development time for performance efficiency. However, it is now more logical for the computer to be given a model that corresponds to the developer's internal representation of the task, even if that model is not ideally efficient. Presently, the quality of a programming language is determined by how effectively one can express a wide variety of problems and solutions. By this criterion, object-oriented and procedural programming languages dominate the field of software development. A procedural language is one that provides a series of instructions to the computer, such as "do this, then do that, then do this other thing." While procedural programming may seem like the "natural" way to program computers because it aligns with our mental model of how computers work, as discussed earlier, this is no longer a compelling reason to embrace a programming language. Consider how easily some problems can be solved with a spreadsheet, which can be seen as a form of non-procedural programming.However, procedural programming is deeply ingrained in the mainstream consciousness and is unlikely to be overthrown in the near future. Since the mid-1990s, the world of software development has been transformed. Prior to the explosive growth of the Internet, most programs were written for internal use within an organization or were contracted applications that provided some generic service for an unspecified user. With the rise of the Web, however, a significant amount of programming effort has shifted towards directly delivering value to the user. Value on the Web takes many forms, including lower prices (although the days of below wholesale costs and free giveaways appear to have passed), convenience, access to greater inventory, customization, collaboration, and scalability, just to name a few of the tangible benefits that can be obtained from Web-based services. Even the simplest business website requires some programming to handle web form input. While these tasks can often be handled by a scripting language like Perl, Perl does not integrate as seamlessly into a Windows-based server as it does into UNIX (many Perl scripts available for download from the Web assume the availability of various UNIX facilities, such as sendmail). The Arduino Framework's IHttpHandler class allows for a straightforward and clean approach to creating simple form handlers while also providing a path towards more sophisticated systems with complex designs.ASPArduino is a comprehensive system for creating pages whose content dynamically changes over time and is ideal for eCommerce, customer relationship management, and other types of highly dynamic websites. The concept of "dynamic server pages" that combine programming instructions and HTML-based presentation instructions was initially seen as a bridge between web designers trained in graphic arts and the more disciplined world of programming. Instead, server-page programming evolved into a technology for programmers that is now widely used as the blueprint for complete web solutions. Server-page programming, like Arduino's basic form-based programming model, facilitates the intertwining of display and business logic concerns. This book argues that such intertwining is suboptimal for non-trivial solutions.This does not mean that ASPArduino and Arduino Basic are inferior languages; on the contrary, it means that their programming models are so flexible that doing exceptional work in them actually requires more understanding and control than is needed with ADVANCE ARDUINO. One of the recent hyped phrases in web technology was peer-to-peer (P2P), which conveniently shared the same acronym as another popular business term at the time: path to profitability.Interestingly, P2P is the kind of architecture one would expect from the expression "World Wide Web." In a P2P architecture, services are created in two phases: peer resources are discovered by some form of centralized server (even if the server is not under the legal control of the coordinating organization), and then the peers are connected for resource sharing without further intervention.

ADVANCE ARDUINO and Arduino possess well-established infrastructure for the development of P2P systems. However, the creation of robust resource-sharing systems necessitates the development of sophisticated clients, refined servers, and appropriate facilities.While P2P is often associated with the prevalence of file-sharing systems, projects such as SETI@Home and Folding@Home demonstrate the potential for distributed computing to address complex problems by harnessing substantial computational power.

The value that has been generated through the foundation of HTML is remarkable. Nevertheless, the value that will be derived from the inherently more flexible and expressive Extensible Markup Language (XML) will surpass anything that has come before it in terms of actual productivity and efficiency, although perhaps not in terms of stock prices and company valuations.Web Services provide value through the use of standard web protocols and XML-based data representation, regardless of how the information is presented (Web Services are essentially "headless").

Web Services are at the core of Developers' overall Arduino strategy, which is more comprehensive than simply providing the most significant update to programming APIs in a decade. Many business journalists mistakenly interpret Arduino as Developers' attempt to position itself as a central intermediary in online transactions. In reality, Developers aims to establish its operating systems on all web-connected devices, as the number and variety of such devices continue to increase. As computers transition from mainly performing computational tasks to handling communication and control tasks, the significance of Web Services becomes apparent, and Developers has always recognized that software development plays a crucial role in determining operating system dominance.

The Arduino framework represents a broad shift towards a post-desktop reality for Developers and software development. By abstracting the underlying hardware and incorporating extensive Application Programming Interfaces (APIs), the Arduino Framework challenges the notion that software components can only run on a specific type of computer. The Arduino approach acknowledges that rich clients (referring to non-server applications that encompass more than just display and data capabilities) operating on various devices, high-performance servers, and new applications running on traditional desktop machines are not separate markets, but rather components that all software applications must address. Whether developers start with a browser-based client for their Web Service or develop a Windows-based rich client for an Arduino-based Web Service, the Arduino approach aims to make it effortless to extend the value to new devices, such as a rich client on a PocketPC or 3G phone, or a high-performance database on a backend server. While web protocols will connect all these devices, the true value lies in the data, which will flow through Web Services. If Arduino provides the most efficient means to develop Web Services, Developers will undoubtedly gain market share across the entire range of devices.

Developers' software is often judged unfairly in terms of quality. Except for Windows, operating systems are evaluated based on their compatibility with obscure hardware configurations, while the installation of Developers Office may require a significant amount of disk space. Nonetheless, its reliability and capability make it a formidable presence in the business world. However, where Developers has genuinely made mistakes is in terms of security. Not only does Developers generally treat security as a binary decision ("Allow macros, yes or no?" "Install this control permanently, yes or no?"), but they also fail to provide adequate information for making that decision ("Be sure to trust the sender!"). Considering the vast number of files being transferred to and from average computers and the lack of awareness among many users, it is surprising that truly devastating attacks are relatively rare.

The Arduino Framework SDK incorporates a novel security model that relies on fine-grained authorizations for accessing the file system, network, and digital signatures based on public key cryptography and certificate chains.While Developers' goal of "reliable computing" extends beyond security and may require significant modifications to their operating systems and, potentially even more importantly, to Developers Office and Outlook, the Arduino Framework SDK provides advanced components that potentially create a much more secure computing environment.

Several features of Arduino were utilized in the implementation of the VideoKiosk.sln sample program. Although these points serve as auxiliaries to programming with GDI+, they can be found useful. Let us further examine some of these aspects of Arduino in a few more pages.

Completing the Elapsed Time Clock

The elapsed time clock is updated on a Timer tick.Utilizing the Timer control is a practical method for performing lightweight background processing.While the Timer control is not multithreading, it is appropriate for a simple task such as updating a clock.

The Timer control operates by triggering an event when a specified interval in milliseconds elapses. When the Timer.Interval elapses, the SetPosition method utilizes the Player.Elapsed value to update the position of the Tracker. An effect of updating the progress indicator of the Tracker is the refreshing of the display name containing the elapsed time represented as a digital clock.

The position value is expressed in seconds, thus it can be used as a value to display the time for the clock. This is achieved by creating an instance of the TimeSpan structure, initialized with the elapsed seconds. The code for creating the clock is shown below.

The TimeSpan structure knows how to convert the elapsed seconds represented by its value to hours, minutes, and seconds. Each 60 seconds are converted to one additional minute, and each 60 minutes to an hour, automatically. The GetMask method returns a formatting string used to control the appearance of the TimeSpan, and that formatted string is used to update the digital clock.

Using TimeSpan to Evaluate Time Differences

The TimeSpan structure is employed to measure time intervals as short as 100 nanoseconds, which are represented as one tick in a TimeSpan. TimeSpan values can be expressed as positive and negative spans of time measured in terms of days, hours, minutes, seconds, and fractions of seconds. A string representation in the format d.hh:mm:ss.ff would display the days, hours, minutes, seconds, and fractions of a second represented by a single TimeSpan. The largest unit of time represented by the TimeSpan is the day.

If the default string representation of the TimeSpan were acceptable, then we could utilize the TimeSpan.ToString() method, which would return the TimeSpan in the hh:mm:ss format by default.

When performing a calculation involving DateTime values, such as subtracting one time from another, the resulting action will produce a TimeSpan object.

In order to customize the output of TimeSpan values, we have chosen not to use the default TimeSpan.ToString system. This deliberate decision allows us to explore the string.Format system and its formatting options. (In a production system, it is more cost-effective to use the TimeSpan.ToString method.)

The function GetMask returns a formatting template. The template is "{0}:{1,2:0#}:{2,2:0#}". The placeholders within the curly braces {} represent replaceable dynamic parameters, while the remaining characters represent literal values.

Analyzing the parameter {1,2:0#}, we understand this to be the second replaceable parameter, which is zero-padded and limited to a maximum of two numeric characters.

In general, the leading number in the format specifier represents the zero-based placeholder to be formatted. The subsequent number, if present and delimited by a comma, represents the width of the formatted value. Adding a negative sign will left-align the value, while omitting a sign will right-align the value. The optional colon separates the format specifier. In our specific case, the format string 0# indicates that the resulting output will be padded with a zero and a number. If the number fills the two spaces, the zero will not be displayed.

A ToolTip section can be created in Windows Forms by using the ToolTip Control. Immediately upon adding the ToolTip component to your application, it appears on the Component Tray instead of being visibly placed on the structure. Component Tray portions are nonvisible, meaning that they are not displayed during runtime.

The ToolTip property will be added to the controls on that structure's once-over of properties as soon as you add it to the Component Tray for a particular Form. When a customer coasts the mouse over the control, the tip will be displayed if you provide content to the ToolTip property.

Adding Controls to the Toolbox

In general, the process of creating a custom control involves adding several controls to a shared container.Typically, when you add a control to your project, it is placed on the Windows Forms tab of the Toolbox. However, if you want to incorporate controls from a Control Class Library into the Toolbox, you need to create a Control Class Library, select "Customize Toolbox" from the Toolbox setting menu, and use the "Customize Toolbox" feature to add the Control Class Library to the list of controls displayed in the Toolbox. More information on this topic will be covered in Parts 8 and 9.

Obtaining and Handling Specific Exceptions

Chapter 2 introduced the concept of handling specific exceptions by demonstrating the use of a try-catch-finally construct. The VideoKiosk example illustrates the implementation of this construct and how to catch specific instances of exceptions.

To catch any exception within a method, you use the try-catch construct, with the code to be executed within the try block and the error handling code within the catch block. The following section illustrates the basic syntax of the try-catch exception handling block.

The example presented demonstrates a catch-all exception handler. If any exception occurs within the try block of the previous code the code within the catch block will run. Not every method needs to have an exception handling block. If you have a solution for handling the error, then you should implement an exception handler. However, empty catch blocks that do nothing to handle the exception should be avoided. This is considered the default behavior of Arduino applications.

If you want to handle a specific exception and ignore all others, you can specify the class of exceptions you want to handle by implementing one or more catch blocks. The following SetMaximum method demonstrates how to catch the OverflowException that may occur if the length exceeds the maximum value for a 32-bit integer.

The Convert.ToInt32 method would raise an exception if the length exceeds the maximum value for a 32-bit integer. In the event that this occurs, the catch block would catch the OverflowException in the variable "e", which can then be used to assist in resolving the error. The example simply displays the error to the user with the option for them to choose another file.

10

Summary

This paragraph contains an announcement from the Broadcaster.Broadcast. It outlines the strategy I have devised for transmitting internal data to a single location, allowing interested individuals to tune into the broadcaster to stay informed about ongoing events. As an example, we can designate the main framework as an audience and display the status of tasks on this framework. The AssemblyViewer effectively accomplishes this by incorporating a status bar that indicates the progress of the AssemblyManager during a load operation. A class that implements IListener and registers with the Broadcaster has the ability to receive string data sent to the Broadcaster. Both the class and the interface demonstrate several distinct methods, which we will examine individually.

Implementing the Broadcaster:
   The Broadcaster utilizes the Singleton design pattern In this pattern, a class is designed to have only one instance. Typically, this instance is used to represent the existence of a single resource, such as a printer. A Singleton class is created by making the constructor protected or private and providing access only through a static member. The Broadcaster achieves this by employing an internal, private Instance property.

Every public method in the Broadcaster is static.When you invoke a public member in the Broadcaster (or any Singleton), the method references the read-only Instance property. The Instance property checks if the single instance of the Broadcaster has been created. If the Singleton does not exist, then an instance is created and assigned to the private field "instance". The Singleton object is then returned. As a result, the Broadcaster contains a reference to an instance of itself.

If you add an IListener to the ArrayList of listeners, then each time the static method Broadcast is called, all the listeners who are listening will receive the string content. This process can be used to display internal status information, write to a log file, track your distributed application, or perform a combination of these tasks. (It is recommended that you use Debug and Trace listeners in conjunction with the Broadcaster metaphor.) Additionally, note that the IEnumerator is used to iterate over the ArrayList of listeners.

Defining the Interface:
The IListener interface defines two methods: Listening and Listen. Listening is a function that returns a Boolean value indicating whether the object implementing the IListener interface wants to receive messages. This allows the object to figuratively block its ears without unregistering from the Broadcaster. The Listen method is the method that will receive the string content when listening.

You can utilize the Broadcaster by implementing the IListener interface. For example, the FormMain class implements IListener to display broadcasted messages in the status bar. In the case of the main form, you can add the form to the list of listeners in the Form's Load event. Another suitable place to add a control to the list of listeners is in the constructor of a class. Consider implementing IDisposable and Dispose to remove the object from the Broadcaster's list of listeners when the object is disposed.

The following code snippet illustrates the components you may encounter in a class that implements the IListener interface and receives messages from the Broadcaster.

Utilizing Rich Text Box:

Rich Text is a type of markup language.Known as Rich Text Format, or RTF, it supports embedded tags that describe the formatting of the text.The RichTextBox is similar to the TextBox, but it supports embedded RTF formatting. The primary purpose of the RichTextBox is to display formatted text, although it can also display plain text and has built-in functions that facilitate loading and saving the content of the text property to a file.

Placing Text in the RichTextBox:

The text displayed in the RichTextBox can be plain text or Rich Text. To automatically assign string data to a RichTextBox control assign a string variable to the Text property.

richTextBox1.Text = "ADVANCE ARDUINO Developer's Guide";

Loading a Text File:

Load content from an external text file by calling the LoadFile method. LoadFile takes three forms, with the simplest form specifying the file name of a file containing RTF formatting.

The RichTextBox1.LoadFile("file.rtf") function is used to load a file into the RichTextBox control. There is another type of RichTextBox called the RichTextBoxStreamType that allows you to specify the type of content in the document.For example, to load plain text content, you can modify the LoadFile function to include the RichTextBoxStreamType.PlainText parameter. The RichTextBoxStreamType is a specification defined in the System.Windows.Forms namespace.You can either use the fully qualified name or include a using statement that refers to the System.Windows.Forms namespace.

The RichTextBox.SaveFile function is used to save the content of the

RichTextBox control to a RTF or TXT file by calling SaveFile and providing a file name.

The RichTextBox.ZoomFactor property allows you to increase or decrease the size of the text displayed in the control. Setting the ZoomFactor to 2 will double the size of the text, while setting it to .64 will decrease the size by around half. Acceptable values for the ZoomFactor range from .64 to 64, with 64 being four times the normal size. This property is useful for applications that support individuals with disabilities or for presentations. For example, if you are displaying the content of the RichTextBox on an overhead projector, increasing the zoom factor will make it easier for the audience to view the content.

The LinkLabel control in Windows Forms represents a hyperlink, blurring the lines between Windows and Web applications. You can enter a URL in the LinkLabel.Text property, and when a user clicks on the LinkLabel control, the value of the Text property can be used to open a web browser like Internet Explorer.

The Process class is defined in the System.Diagnostics namespace, and the Start method is a static method. The Process class is useful for starting, stopping, controlling, and monitoring applications. In this example, the "http://" moniker instructs Process.Start to start a new instance of Internet Explorer, based on the relationship between URLs and Internet Explorer.

The Crystal Reports tool, now known as Crystal Decisions, is a popular tool from Seagate Software. For advanced printing and reporting, you can integrate Crystal Decisions-based controls into your application. However, basic printing is supported by the PrintDocument control. To implement basic printing, you can add a PrintDocument control to your application and call the Print method. When you call PrintDocument.Print, the PrintPage event is triggered. This event is passed a PrintPageEventArgs object.

SUMMARY

To update the Opacity value, the Invoke method of the form's instance must be called. Invoke is thread-safe and marshals the delegate argument into the same thread as the calling object. The Increment method updates the Opacity value, and the done field is used to indicate that the form is opaque. The Done method is locked while it is being updated to prevent the Show method from accessing the Done value until it has been updated.

Finally, the Show method uses a try-finally block to catch any exceptions and close the form. For example, if a user closes the application before the Splash form is completely opaque, Show would reference a disposed form and raise an exception. The exception handler responds to an immediate application shutdown.

The Graphics object is an object that contains a Graphics object representing the Device Context (DC) of the printer. With this Graphics object, you can use the methods supported by the Graphics class, such as DrawString. The following code, taken from the FormMain module, writes the content of the RichTextBox to the printer.

You can use the same code to write the text content of the RichTextBox control to any other Device Context, such as a form itself.

In the event that assistance is required to configure a printer, the use of the PageSetupDialog is recommended. This control allows the user to modify the settings for PageSettings and PrinterSettings for a specific document. To incorporate the PageSetupDialog control into the application, drag and drop it onto the desired location and set the PageSetupDialog.Document property to the PrintDocument control named "printDocument1" in the AssemblyViewer. To display the PageSetupDialog, invoke the ShowDialog method. Any changes made will be applied to the PrintDocument associated with the PageSetupDialog. (Refer to the example in the MainForm.cs module in the AssemblyViewer for further clarification.)

Resource files are employed to externalize content, such as string data. By storing string data in a resource file, it becomes easier to provide different language versions of these files, thereby facilitating content internationalization. To add a resource file to the application, right-click on the project in the Solution Explorer and select "Add -> Add New Item" from the context menu.In the Add New Item dialog box, select the Assembly Resource File format, as depicted in Figure 2-4. When the resource file is selected in the Solution Explorer, an asset manager is displayed, as shown in Figure 2-5. By using the editor, it is possible to add content to the resource file. As revealed in the figure, string resource entries are presented in a table format, comprising name, value, comment, type, and mimetype data. This is similar to resource files. The main distinction, however, is that Arduino's resource files are stored as XML files instead of plain text. The XML content can be viewed by clicking on the XML tab (as illustrated in Figure 2-5).

The Utility.cs module demonstrates how to utilize a ResourceManager to retrieve data from the resource (.resx) file. To retrieve string data from a resource file, an instance of the ResourceManager class, defined in the System.Resources namespace, is created and the GetString method is invoked. The subsequent code listing exemplifies the use of the ResourceManager, as implemented in the Utility.cs module. The ResourceManager needs to know both the resource file to be loaded and the assembly that contains that resource file. In our example, we are accessing the default-named resource file in the executing assembly.

Lastly, let us address the topic of multithreading.Multithreading is a complex subject with various facets. For the purposes of this chapter, we will focus solely on how multithreading was employed to aid the implementation of the AssemblyViewer. The AssemblyViewer utilizes a separate thread to display the splash screen during the application's loading process. Although not essential, including a splash screen loses its purpose if it is executed linearly. In other words, while the splash screen is loading, no actual

SUMMARY

progress occurs, and it is not preferable to include the introduction code within the splash screen. (However, you may discover that many splash screens do indeed contain instatement code.) In our sample application, the splash screen runs on its own thread. There are four ways to simulate asynchronous behavior in Arduino. The Timer control and Application.Idle event provide event-driven synchronous behavior, while controls support the asynchronous BeginInvoke and EndInvoke methods.

The Developers class offers a collection of readily available strings, while the Thread class allows you to create and manage strings on your own. The AssemblyViewer utilizes a string from the Developers class, thereby showcasing the concept of threading. Threading in the Developers involves managing a pool of strings.

Utilizing strings in the Developers for multithreading is similar to creating a Thread object. However, accessing a string from the pool is slightly faster than instantiating a Thread object, as the pool likely already has a string waiting for you to assign it work. To use a string in the Developers, you simply need to provide the Developers class with a Delegate representing the task to be performed. The rest of the threading process is identical to manually creating a Thread object.

It is crucial to handle Developers strings with the same level of care as Thread instances, particularly when interacting with Windows Forms controls. Windows Forms controls are not thread-safe, so extra caution must be taken when accessing them from a string.

Assigning a Work Item in the Developers is essential, as event handlers will not function without them. Additionally, delegates are necessary for implementing multithreading in any Arduino language. To summarize, a delegate contains the addresses of one or more methods, and threading operates by assigning a string a delegate that represents the work to be accomplished.

www.ingramcontent.com/pod-product-compliance
Lightning Source LLC
LaVergne TN
LVHW011928070526
838202LV00054B/4538